民航英语特色课程系列教材

Aviation English for Flight Crew
飞行英语阅读
（第2版）

主　编　李明良
副主编　申卫华　蒲建君　陈华妮
　　　　任晓军　朱　敏　何　莉
审　校　刘德仲

U0205569

西南交通大学出版社
·成　都·

图书在版编目（CIP）数据

飞行英语阅读 / 李明良主编. —2 版. —成都：
西南交通大学出版社，2017.6（2024.7 重印）
民航英语特色课程系列教材
ISBN 978-7-5643-5470-1

Ⅰ. ①飞… Ⅱ. ①李… Ⅲ. ①民用航空－英语－阅读
教学－高等学校－教材 Ⅳ. ①V2

中国版本图书馆 CIP 数据核字（2017）第 124698 号

民航英语特色课程系列教材

Aviation English for Flight Crew
飞行英语阅读
（第 2 版）

主编 李明良

责 任 编 辑	赵玉婷
封 面 设 计	何东琳设计工作室
出 版 发 行	西南交通大学出版社
	（四川省成都市二环路北一段 111 号
	西南交通大学创新大厦 21 楼）
发 行 部 电 话	028-87600564　028-87600533
邮 政 编 码	610031
网 址	http://www.xnjdcbs.com
印 刷	成都蓉军广告印务有限责任公司
成 品 尺 寸	185 mm×260 mm
印 张	18
字 数	513 千
版 次	2017 年 6 月第 2 版
印 次	2024 年 7 月第 14 次
书 号	ISBN 978-7-5643-5470-1
定 价	48.00 元

编写说明

  《飞行英语阅读》是根据国际民航组织飞行员英语语言能力要求，结合国内航空公司对于飞行员英语水平的要求和实际飞行、训练需要，形成的具有飞行特色和时代特色的专业英语背景基础知识阅读教材。该教材由书本、光盘、网络等多种载体构成，融汇了国内现行大学英语阅读教材、其他文科教材和之前的民航英语阅读教材的编写理念和模式，旨在帮助民航飞行技术专业大学生奠定民航英语基础知识，培养学生的民航英语综合应用能力。该教材是国内第一本专门提高飞行技术专业大学生民航英语能力的阅读教材，标志着国内民航飞行员专业英语阅读教学迈出探索特色课程建设的第一步。

  本教材系飞行英语系列教材之基础教材。全书共 4 个单元，包括 16 课，供已完成大学英语学习的飞行技术专业学生使用。每课包括 Text A 和 Text B 两个部分，各部分包含相关配套练习。

  一、《飞行英语阅读》的选材

  《飞行英语阅读》在选材上立足于飞行技术专业大学生的专业英语学习阶段的语言基础和专业英语基础语言知识学习的需要，同时兼顾国际民航组织飞行员英语水平要求及标准、飞行员实际训练和工作的需要。本书强调"以学生自主学习为中心，教师适时引导"的教学原则。在选材上注重基础性、实用性、全面性、系统性和前瞻性。本书选材涉及飞行知识的四个基本方面：人、机、程序和环境。每个单元内容自成体系，Text A（800～900 字）和 Text B（约 600 字）取材具有典型性，每课力求专业英语知识的系统性，以引导读者在课外进行相关英语语言知识的学习。

  二、《飞行英语阅读》的主要内容

  《飞行英语阅读》每课开头均有 Preview。Preview 言简意赅地点明本课的中心内容。它是本课的导读篇，对于学生课外自主学习具有导示作用。

  每课包括 Text A 和 Text B 两个部分。Text A 包括 Warming-up Activities, Passage, New Words, Expressions 和 Notes, Exercises 包括以下几个部分：课文术语解释、理解问题、相关重要题材短文朗读、相关题材短文的选词填空、阅读理解和补充阅读。这种建构形式有别于国内其他阅读教材阅读文本的单一性，其形式和内容以民航飞行英语涉及内容的特点为基础。

  Text B 的课文便于学生课外学习，作文练习引导学生运用所学的相关民航英语知识进行写作练习。这便于教师检查，同时为学生提高口语能力奠定基础。

  三、《飞行英语阅读》的语言练习项目

  《飞行英语阅读》在复杂、庞大的飞行英语知识系统中构建了基本合理的课目和单元构架。教材综合国内时新阅读教材的单元练习体例，继承了国内经典阅读教材的有效阅读练习形式，结合飞行英语知识需要，形成独特而适用的主题—任务性练习体系结构，注重内容全面性、语言知识的习得和语言的实际应用等方面。

术语解释侧重学生应用自己的英语语言知识和专业英语知识对 Text A 中的一些重要术语进行口头或书面的表达练习，同时也可以检查学生对于课文的学习和掌握情况。

课文理解问题的问答有助于学生提高就常见、具体专业性问题的口头表达能力。朗读练习强调语音语调，让学生熟悉重要话题的短文（200字）的语言知识，同时促使学生重视语音语调的标准性。

完型填空和阅读理解题型是两种经久不衰的阅读练习形式，它们能在很大程度上检查学生的语篇理解能力，更重要的是能让学生在这种练习中应用已有专业英语知识分析问题、解决问题，从而提高学生专业英语语言应用能力。这两种练习都以短文文本为对象，既能丰富学生的专业英语背景知识，满足国际民航组织对于飞行人员英语语言能力要求的需要，又传承了国内流行的阅读理解理念和练习形式，强调了民航英语语言学习和应用这一基本理念。

每篇补充阅读文章约500字左右，重要生词都标注了语义，便于学生自主补充有关单元主题的阅读内容和丰富学生的背景知识。

Text B 采用写作练习，让学生应用本课所学知识，结合本部分课文提供的专业英语知识，就某一给定话题进行写作练习，增进相关民航英语知识习得，提高专业话题的记叙、阐述和描述等英文表达能力，在国际民航组织飞行员英语语言能力要求和标准方面具有较强的针对性。

四、《飞行英语阅读》的阅读量和词汇量

《飞行英语阅读》全书共85 000词，其中精读阅读量约24 000词，练习阅读量约40 000词。

五、《飞行英语阅读》的编者

《飞行英语阅读》由中国民航飞行学院外国语学院李明良担任主编，编写本教材的多数内容。中国民航飞行学院外国语学院蒲建君、申卫华、陈华妮和任晓军等老师分别参加编写4～6课、7～9课、10～13课和14～16课的部分内容。感谢朱敏老师为教材样式设计与修改等提供的宝贵的建设性意见和做的大量工作。

《飞行英语阅读》由中国民航飞行学院外国语学院民航英语专家刘德仲审定全稿。

《飞行英语阅读》的设计、编写和制作是一个研究现行各种英文阅读教材、探索具有中国民航飞行员特色的新时代民航英语阅读教材的过程。由于编者水平所限，教材中不乏不足之处，诚挚欢迎国内 ESP 专家、民航界专家、民航英语专家以及使用本教材的广大师生给予批评和指正。

《飞行英语阅读》教材编写组

2013年8月

前　言

对于飞行员和管制员来讲，无线电陆空通话语言不仅是交流和空地对话的工具，而且是保证航空安全的重要因素之一。

早在1998年，针对飞行人员和管制人员语言能力不足而造成的飞行事故，国际民航组织第32届大会在A32-16决议中就敦促ICAO理事会对ICAO附件1和附件10进行修订，要求参与国际运行的飞行人员和管制员应该具有足够的保证航空安全的英语通讯能力。2003年3月，ICAO理事会推出了针对通讯语言能力的标准和建议措施（SARPs），并对与语言熟练程度相关的附件1、6、10、11和PANS-ATM（空中航行服务程序-空管）进行了修订。2004年，国际民航组织出版了《ICAO语言熟练程度要求执行手册》（ICAO DOC 9835），对与语言熟练程度要求相关的培训和测试做出了具体要求。随后，包括中国在内的很多成员国开始开发研制本国的测试系统或同时为其他国家展开测试服务。

2007年，国际民航组织大会通过了A36-11决议《用于无线电通讯的英语语言熟练程度》，要求各成员国采用全球统一的语言测试标准。在此决议的基础上，国际民航组织先后发布了《促进全球统一化的语言测验标准》（318号通告，2009年6月）和《航空英语培训方案指导原则》（323号通告，2010年）。《促进全球统一化的语言测验标准》要求各成员国根据国际民航组织的语言能力要求处理对候选人的测验事宜，并就航空语言测验方案的制作或选择提供建议的标准；《航空英语培训方案指导原则》是应各主管当局、运营人和服务提供者对制定英语语言培训更详细的指导的请求而与国际民用航空英语协会（ICAEA）联手制作的，以便能有效地实施附件1《人员执照的颁发》中所载的语言能力要求。2010年9月，国际民航组织出版了《ICAO语言熟练程度要求执行手册》（ICAO DOC 9835，第二版)，对原有的有关民航语言测试和培训等内容做了较大的修改，对标准和原则做了更加细致的说明，并结合了318号和323号通告对测试和培训的操作标准。

针对国际民航组织的标准，中国民用航空局于2006年启动了针对飞行人员的英语等级执照签注考试，2007年开始PEPEC考试系统的立项和研制，并于2008年开始使用具有中国民航自主知识产权的PEPEC考试系统。中国民航局空中交通管理局于2005年委托中国民航飞行学院开始进行中国民航空管人员英语等级考试系统（AETS）的开发工作，并于2007年12月开始AETS的考试。

作为世界上最大的航空训练机构，中国民航飞行学院是中国民航培养飞行员和空中交通管制员的主力院校。飞行学院的毕业生不仅要满足法规所规定的执照签注要求，还要满足国际民航组织所要求的国际运行所需要的语言熟练程度执行标准。中国民航飞行员和管制员语言能力的提高不仅是保证安全的需要，同时也是中国民航从民航大国迈向民航强国所必须完成的任务。

中国民航飞行学院外国语学院就是在这样的背景下提出了"民航英语特色课程系列教材"的编写。学院和民航局给予了大力支持，该系列教材的建设成为我校"以英语教学改革为龙头，推动全面教学改革"的教改项目和民航局资助的"国际民航组织英语语言课程建设"项目。

本系列教材共六本：《飞行英语阅读》（李明良主编）《飞行英语听说》（陈方主编）《飞行英语口语》（陈方主编）《飞行英语无线电陆空通话高级教程》（申卫华主编）《管制英语阅读》（郑丽主编）《管制英语无线电通话》（何均洪主编）。该系列教材可用于飞行技术专业和空管专业讲授民航英语的高段课程，也可作为航空公司或空中交通管理局培训部的培训教材。

民航英语特色课程系列教材旨在提高飞行技术专业学生和空管专业学生应用英语语言的综合能力、使用英语进行通信的能力，以及针对一些特殊情况的处理能力。本系列教材从六个方面对学员进行语言能力的培养（语言、结构、词汇、理解、流利程度、应对能力），全面提高学员的交际能力和应用能力。教材的编写全面贯彻国际民航组织（ICAO）新语言标准。

民航英语特色课程系列教材中所编制的情景的主要依据是国际民航组织《语言熟练程度要求执行手册》（ICAO DOC 9835，2010）、中国民航飞行人员英语等级考试（PEPEC）大纲和民航管制员英语考试（AETS）大纲中所列出的事件和范围，以及近年来世界范围内一些典型的航空事故和事件。

民航英语特色课程系列教材项目总负责人为中国民航飞行学院陈布科副校长，参与编写人员有外国语学院教师、空管学院教师、继续教育学院教师、空乘学院教师、分院民航英语教师、飞行教员、航空公司飞行员，同时由资深教师、资深飞行员和我院外籍教师审定。

民航英语特色课程系列教材编写组
2013 年 4 月

Contents

Unit One
Air Crew

Lesson 1

Air Crew

Preview

> The pilot, a vital asset to the survival of any airlines, needs to know what elements constitute a qualified pilot. Text A serves to offer some fundamental knowledge concerning duties and awareness of captains, F/Os, PICs, student pilots, and flight instructors. Text B is centered on the job of cabin crew.

Text A

Warming-up Activities

Picture Description

Please describe the following picture and be prepared to answer some questions.

Relevant Questions

1. Which airline will you work with after graduation?
2. Why do you want to be a pilot?
3. What should you often do to be a good pilot?
4. What is your long-term plan to be a qualified captain?

Captain and F/O Duties

There are usually 2-3 flight crew members and 1-3 flight attendants aboard an airliner. In the flight deck are the Captain, Co-pilot and flight engineer. When there are only two flight crew members, to reduce costs there's no flight engineer. The Captain is the Pilot in Command (PIC) who has the final authority of all decisions and all responsibilities rest on his shoulders. The Co-pilot assists the Captain and does things like calculating fuel consumption, weight and balance, navigation, etc. He is Second in Command (SIC). The Flight Engineer helps reduce the workload of the Captain and Co-pilot. Some of his duties may include calculating fuel consumption rate, weight and balance, and communicating with the cabin crew.

The Pilot in Command (PIC) of an aircraft is the person aboard the aircraft who is ultimately responsible for its operation and safety during flight. According to ICAO, the Pilot in Command is responsible for operating an airplane in accordance with rules of the air, and has final authority as to the disposition of the airplane while in command. This would be the "captain" in a typical two- or three-pilot flight crew, or "pilot" if there is only one certified and qualified pilot at the controls of an aircraft. The PIC must be certified to operate the aircraft for the specific flight and flight conditions, but need not be actually controlling the

aircraft at any given moment. The PIC is the person legally in charge of the aircraft and its flight safety and operation, and would normally be the primary person liable for a violation of any flight rule.

According to FAA, the PIC is responsible for the operation and safety of the aircraft during flight time, which means the total time from the moment an aircraft first moves under its own power for the purpose of taking off until the moment it comes to rest at the end of the flight. This would normally include taxiing, which involves the ground operation to and from the runway. But it is legal for a mechanic or other person to taxi an aircraft on the ground for the purpose of moving it from one spot to another without a pilot's license.

As the FAA puts it, the pilot in command of an aircraft is directly responsible for, and is the final authority as to, the operation of that aircraft. In an in-flight emergency requiring immediate action, the pilot in command may deviate from any rule of this part to the extent required to meet that emergency. Each pilot in command who deviates from a rule of this section shall, upon the request of the administrator, send a written report of that deviation to the administrator.

Especially interesting is FAR 91.3, which empowers the PIC to override any other regulation in an emergency, to take the safest course of action at his/her sole discretion. It essentially gives the PIC the final authority in any situation involving the safety of a flight, irrespective of any other laws or regulations.In commercial aviation, the first officer is the second pilot (or co-pilot) of an aircraft. The first officer is second-in-command of the aircraft. In the event of incapacitation of the captain, the first officer will take on the duties of the PIC.Control of the aircraft is normally shared equally between the first officer and the captain, with one pilot being the "Pilot Flying"(PF), and the other the "Pilot Not Flying" (PNF), or "Pilot Monitoring" (PM), for each flight. Even when the first officer is the flying pilot, however, the captain remains ultimately responsible for the aircraft, its passengers, and the crew. In typical day-to-day operations, the essential job tasks remain fairly equal.

Because many airlines promote by seniority only within their own company, the first officer may at times have more flight experience than the captain, in that they may have experience from other airlines or the military. Traditionally, the first officer sits on the right-hand side of a fixed-wing aircraft and the left-hand side of a helicopter. (686 words)

NEW WORDS

| crew | /kruː/ | n. | 人员，组员 |
| airliner | /ˈeəˌlainə/ | n. | 公司班机 |

captain	/'kæptin/	n.	机长
command	/kə'mɑ:nd/	n. & v.	指挥；指令
authority	/ɔ:'θɒriti/	n.	权威，权限
fuel	/fju:əl/	n.	燃油
navigation	/ˌnævi'geiʃən/	n.	导航；领航
communicate	/kə'mju:nikeit/	v.	交流；通讯
briefing	/bri:fiŋ/	n.	讲述；讲评
emergency	/i'mə:dʒənsi/	n.	紧急
procedure	/prə'si:dʒə/	n.	程序
evacuation	/iˌvækju'eiʃən/	n.	撤离
aircraft	/'eəkrɑ:ft/	n.	飞机；航空器
operation	/ˌɒpə'reiʃən/	n.	运转，工作，操作
disposition	/ˌdispə'ziʃən/	n.	处置，处理
certify	/'sə:tifai/	v.	认证；认可；证明
manipulate	/mə'nipjuleit/	v.	操纵
taxi	/'tæksi/	n. & v.	滑行
runway	/'rʌnwei/	n.	跑道
mechanic	/mi'kænik/	n.	机务员，机械人员
license	/'laisəns/	n. & v.	执照；授证
deviate	/'di:vieit/	v.	偏离(from)
administrator	/əd'ministreitə/	n.	管理人
discretion	/dis'kreʃən/	n.	决定
override	/ˌəuvə'raid/	vt.	超控
provision	/prə'viʒn/	n.	条款
justification	/ˌdʒʌstifi'keiʃn/	n.	证明
incapacitation	/ˌinkəpæsi'teiʃn/	n.	失能
airline	/'eəlain/	n.	航空公司
seniority	/ˌsi:ni'ɔ:rəti/	n.	资历
wing	/wiŋ/	n.	机翼
helicopter	/'helikɒptə/	n.	直升机

EXPRESSIONS

flight crew　机组人员　　　　　　Pilot Flying　操纵飞机飞行员
flight attendant　空中乘务员　　　Pilot Not Flying　未操纵飞机的飞行员
flight engineer　飞行机械师　　　　Pilot Monitoring　监控飞行员

Pilot in Command　责任机长　　　　First Officer　副驾驶

NOTES

ICAO	International Civil Aviation Organization	国际民航组织
FAA	Federal Aviation Administration	联邦航空局（美国）
FAR	Federal Aviation Regulations	《联邦航空条例》（美国）

EXERCISES

Comprehension of Text A

I. Describe the given aviation terms from Text A in English.

1. flight deck
2. PIC
3. incapacitation
4. flight time
5. captain
6. PNF

II. Answer the following questions after you have read Text A.

1. What crew members are included in an airplane?

2. What are the duties for different crew members?

3. In what condition will the co-pilot take on the responsibility of commanding an aircraft?

4. Can you tell the differences between PIC, captain and pilot?

5. According to FAA, what are the responsibility and authority of PIC?

6. According to ICAO, what is the Pilot in Command responsible for?

7. Why does FAR give final authority to PIC in an emergency?

8. Should PF always be the Pilot in Command? Give your reasons.

9. Why do some First Officers have more experience than their captains? Can you give an example?

10. Where does the first officer normally sit in a helicopter?

Reading Aloud

III. Read the following paragraph aloud until you can say it in a natural way from your memory.

Learning to fly from a flight instructor （飞行教员）is one of the most exciting and costly

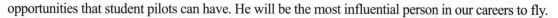

opportunities that student pilots can have. He will be the most influential person in our careers to fly.

Instructors are life savers. Making a mistake in a plane can lead to a lot of problems unless the instructor is qualified. Often, the quality of a pilot, especially in the early stages, is a reflection of the quality of the person who has taught them. During flight training, the instructor will save us and the aircraft from many of mistakes, which are possibly fatal. Therefore, an instructor is supposed to have appropriate experience and knows how to correct the problem, or in extreme cases, to take over and save both of our lives.

Good instructors make great pilots. It is important to have a strict instructor who will work with us in order to reach perfection. A patient instructor can make young pilots feel confident in their abilities. Also, teaching technique is very important. Many instructors prefer to teach in briefing rooms before taking students up into the air. Others bring students up in the air right at the beginning and teach them how to fly by having them fly. (205 words)

Vocabulary

IV. Complete the following short passage by filling the blanks with the words given in the box.

> optimal   airplanes   while   transport   regulations   flight
> turns   functioning   in-flight   communicate

Airline Pilots

Airline pilots 1_____ passengers and cargo（货物）via commercial aircraft. They are responsible for various 2_____ and non-flying duties. Airline pilots are responsible for a variety of tasks related to the safe operation of the 3_____ they are responsible for flying.

Prior to takeoff, airline pilots are responsible for verifying that the instruments（仪表）, controls, engines, and other flight systems are 4_____ the way they should. Monitoring continues throughout the 5_____, and changes are requested and implemented as needed 6_____ en route（航路上）.

On commercial flights, there are two pilots at the helm of the airplane. Each flight is staffed by a captain and a first officer, and the two pilots typically take 7_____ flying different legs（航段）of each trip. They 8_____ with flight dispatchers（签派人员）, air traffic controllers, and meteorologists to select the best route（航线）for their trip, including determining the 9_____ altitude and speed for travel.

Once a flight lands, airline pilots are required to complete records about their journeys in compliance with the 10_____ of both the company for which they work and the Federal Aviation Administration (FAA).

Reading for More

V. Read the following passage and answer the given questions briefly.

To Be a Student Pilot

The first step for you to become a pilot is to get your FAA Medical Certificate prior to starting any flight training. This way, you'll know for sure that your physical condition meets the FAA standards. When you take the physical examination from a FAA Designated Medical Examiner in your area, we recommend that you apply for a Student Pilot / 3rd Class Medical Certificate, since this is the minimum required prior to soloing (单飞).

In order to hold a Student Pilot Certificate, you must successfully complete the necessary ground and flight training covering the required aeronautical (航空) and piloting (驾驶) skills necessary for safe solo flight. You must pass the Pre-Solo Written Test, given by your instructor, prior to your first solo flight. To solo usually takes between 10-20 hours of flight instruction. Once you and your instructor have determined that you're ready, you will be given the opportunity to take the aircraft for your first solo flight! This will undoubtedly be one of the most memorable days you will ever experience!

After your first solo, you'll still continue your flight instruction with your instructor on a regular basis. But now you will complement that instruction by practicing maneuvers on your own to reinforce what you have learned and to gain skills, judgment, and confidence to operate an airplane all by yourself. This solo time is necessary towards the flight experience requirements for your Private Pilot Certificate (私人驾驶执照). As a Student Pilot, you are restricted from carrying passengers, and can only fly in the area or to the airports that your instructor has authorized. (263 words)

Question 1 Why should a pilot have a Medical Certificate before flight training?

Question 2 To hold a Student Pilot Certificate, what tasks are student pilots required to accomplish?

Question 3 According to the passage, what should you do after the first solo?

VI. In this section, there are two passages. After you have read the two passages, there are five questions followed by four choices marked A, B, C and D. Decide which one is the most appropriate answer.

Passage One

Positive Transfer of Control

During flight training, there must always be a clear understanding between the student and flight instructor of who has control of the aircraft. Prior to any dual training flight, a briefing (讲评) should be made that includes the procedure for the exchange of flight controls. The following three-step process for the exchange of flight controls is highly recommended.

When a flight instructor wishes the student to take control of the aircraft, he/she should say to the student, "You have the flight controls." The student should acknowledge immediately by saying, "I have the flight controls." The flight instructor confirms by again saying, "You have the flight controls." Part of the procedure should be a visual check (目视检查) to ensure that the other person actually has the flight controls. When returning the controls to the flight instructor, the student should follow the same procedure the instructor used when giving control to the student. The student should stay on the controls until the instructor says, "I have the flight controls." There should never be any doubt as to who is flying the airplane at any one time. Numerous accidents have occurred due to a lack of communication or misunderstanding as to who actually had control of the aircraft, particularly between students and flight instructors. Establishing the above procedure during initial training (初始训练) will ensure the formation of a very good habit pattern. （226 words）

1. What does the first sentence of the first paragraph mean?

    A. Control of the airplane can never be neglected.

    B. The three steps for exchange of control should be stressed.

    C. It is the instructor who always has the final saying in training.

    D. Preflight briefing is an integral part of any flight training.

2. Besides saying "You have the flight controls", the instructor should _____.

    A. make a briefing   B. acknowledge the transfer

    C. make a visual check   D. keep monitoring systems

3. When receiving the transfer of control, the student pilot should _____.

    A. acknowledge and take control   B. make a visual check

    C. check the flight controls   D. confirm the check procedure

4. The purpose of the three-step procedure is to ensure the aircraft _____.

    A. is controlled by two pilots   B. is controlled at any one time

    C. is following the same procedure   D. is undoubtedly flying on time

5. "A very good habit pattern" refers to _____.

    A. the control of the airplane   B. a specific flight progress

    C. the lack of communication   D. the three-step procedure

Passage Two

Collision Avoidance Awareness

All pilots must be alert to the potential for midair collision and near midair collisions.

The general operating and flight rules in 14 CFR part 91 set forth the concept of "See and Avoid". This concept requires that vigilance (警惕) shall be maintained at all times by each person operating an aircraft, regardless of whether the operation is conducted under instrument flight rules (仪表飞行规则) or visual flight rules (目视飞行规则). Pilots should also keep in mind their responsibility for continuously maintaining a careful lookout regardless of the type of aircraft being flown and the purpose of the flight. Most midair collision accidents and reported near midair collision incidents occur in good VFR weather conditions and during the hours of daylight. Most of these accident/incidents occur within 5 miles of an airport and/or near navigation aids (助航台).

The "See and Avoid" concept relies on knowledge of the limitations of the human eye, and the use of proper visual scanning (扫视) techniques to help compensate for these limitations. The importance of, and the proper techniques for, visual scanning should be taught to a student pilot at the very beginning of flight training. The competent flight instructor should be familiar with the visual scanning and collision avoidance information, pilots' role in collision avoidance, and the Aeronautical Information Manual (航空情报手册).

There are many different types of clearing procedures. Most are centered around the use of clearing turns. The essential idea of the clearing turn is to be certain that the next maneuver (机动) is not going to proceed into another airplane's flight path (飞行轨迹). Some pilot training programs have hard and fast rules, such as requiring two 90° turns in opposite directions before executing any training maneuver. Other types of clearing procedures may be developed by individual flight instructors. Whatever the preferred method, the flight instructor should teach the beginning student an effective clearing procedure and insist on its use. The student pilot should execute the appropriate clearing procedure before all turns and before executing any training maneuver. Proper clearing procedures, combined with proper visual scanning techniques, are the most effective strategy for collision avoidance. (345 words)

6. From this passage the word "clearing" can be replaced by _____.
   A. avoiding　　B. cleaning　　C. limiting　　D. alerting
7. Most midair collision occurred in good weather conditions because _____.
   A. pilots are not familiar with instrument flight
   B. pilots ignore the airplanes' flight characteristics
   C. pilots do not keep looking out in the air
   D. pilots are unaware of the purpose of their flights
8. A flight instructor should teach students to learn visual scanning techniques to _____.
   A. enrich their knowledge of instrument flight

B. know the roles of student pilots

C. compensate for human eye's limitation

D. ensure normal operation of instruments

9. To make a clearing turn, it is important not to _____.

    A. follow hard training programs   B. obey the rules of avoiding traffic

    C. execute another training maneuver   D. invade another airplane's flight path

10. From the end of the passage, it seems obvious that _____.

    A. flight instructors should get well along with student pilots

    B. clearing procedures go with proper visual scanning

    C. effective strategy for collision avoidance is not easy to acquire

    D. "See and Avoid" concept relies only on students themselves

VII. Read the following passage and retell it in your own words.

Role of the Flight Instructor

The flight instructor is the **cornerstone**[1] of aviation safety. The FAA [1] 奠基石 has adopted an operational training concept that places the full responsibility for student training on the authorized flight instructor. In this role, the instructor assumes the total responsibility for training the student pilot in all the knowledge areas and skills necessary to operate safely and competently as a **certificated**[2] pilot in the National Airspace System. This training will [2] 持照的 include airmanship skills, pilot judgment and decision making, and accepted good operating practices.

An FAA certificated flight instructor has to meet broad flying experience requirements, pass knowledge and practical tests, and demonstrate the ability to apply recommended teaching techniques before being certificated. In addition, the flight instructor's certificate must be renewed every 24 months by showing continued success in training pilots, or by satisfactorily completing a flight instructor's refresher course or a practical test designed to upgrade **aeronautical**[3] knowledge, pilot [3] 航空的 proficiency, and teaching techniques.

A pilot training program is dependent on the quality of the ground and flight instruction the student pilot receives. A good flight instructor will have a thorough understanding of the learning process, knowledge of the fundamentals[4] of teaching, and the ability to communicate effectively with [4] 基础 the student pilot.

A good flight instructor will use a **syllabus**[5] and insist on correct [5] 教学大纲
techniques and procedures from the beginning of training so that the student
will develop proper habit patterns. The syllabus should embody the "building
block" method of instruction, in which the student progresses from the
known to the unknown. The course of instruction should be laid out so that
each new maneuver embodies the principles involved in the performance of
those previously undertaken. Consequently, through each new subject
introduced, the student not only learns a new principle or technique, but
broadens his/her application of those previously learned and has his/her
deficiencies[6] in the previous maneuvers emphasized.   [6] 缺陷

The flying habits of the flight instructor, both during flight instruction
and as observed by students when conducting other pilot operations, have a
vital effect on safety. Students consider their flight instructor to be a
paragon[7] of flying proficiency whose flying habits they, consciously or [7] 典范
unconsciously, attempt to imitate. For this reason, a good flight instructor
will strictly observe the safety practices to teach the students. Additionally, a
good flight instructor will carefully observe all regulations and recognized
safety practices during all flight operations.

Generally, the student pilot who enrolls in a pilot training program is
prepared to commit considerable time, effort, and expense in pursuit of a
pilot certificate. The student may tend to judge the effectiveness of the flight
instructor, and the overall success of the pilot training program, solely in
terms of being able to pass the **requisite**[8] FAA practical test. A good flight [8] 必要的
instructor, however, will be able to communicate to the student that
evaluation through practical tests is a mere sampling of pilot ability that is
compressed into a short period of time. The flight instructor's role, however,
is to train the "total" pilot. （500 words）

Text B

Passage Writing

VIII. Read the following passage carefully and write a passage under the given title with no less than 200 words.

Cooperation with Flight Attendants

_____

_____

Flight Attendants

The majority of a flight attendant's duties are safety related. Prior to each flight, flight attendants attend a safety briefing with the pilots and purser. During this briefing they go over safety and emergency checklists, the locations and amounts of emergency equipment and other features specific to that aircraft type. Boarding particulars are verified, such as special needs passengers, small children or VIPs. Weather conditions are discussed including anticipated turbulence. Prior to each flight, a safety check is conducted to ensure all equipment such as lifevests, flashlights and firefighting equipment are on board, in the right quantity, and in proper condition. Any unserviceable or missing items must be reported and rectified prior to takeoff. They must monitor the cabin for any unusual smells or situations and maintain certain precautions such as keeping doors disarmed or open during fuelling on the ground. They assist with the loading of carry-on baggage, checking for weight, size and dangerous goods. They then must do a safety demonstration or monitor passengers as they watch a safety video demonstrating the safety features of the aircraft. They then must secure the cabin ensuring tray tables are stowed, seats are in their upright positions, armrests down and carry-ons stowed correctly and seatbelts fastened prior to takeoff. All the service between boarding and take-off is called Pre-Takeoff Service.

Flight attendants must conduct cabin checks every 20-30 minutes, especially during night flights, to check on the passengers, and listen for any unusual noises or situations. Checks must also be done on the lavatory to ensure the smoke detector hasn't been deactivated, there are no issues with the equipment, nobody having trouble in there or smoking. Regular cockpit checks must be done to ensure the pilot's health and safety. They must respond immediately to call lights dealing with special requests and smaller emergencies including a wide variety of in-flight emergencies that do happen from time to time. During turbulence, crosschecks must be conducted and during severe turbulence all service equipment must also be stowed. Prior to landing all loose items, trays and garbage must be collected and secured along with service and galley equipment. All hot liquids must be disposed of. A final crosscheck must then be completed prior to landing. They must remain aware as the majority of mechanical emergencies occur during takeoff and landing. Upon landing, flight attendants must remain stationed at exits and monitor the airplane and cabin as passengers disembark the plane. They also assist any special needs passengers and small children off the airplane and escort children, while following the proper paperwork and ID process to escort them to the designated person picking them up.

Flight attendants are highly trained for a wide variety of emergencies and how to respond. More frequent situations may include a bleeding nose, illness, small injuries, drunken passengers, and aggressive and anxious passengers. Emergency training includes rejected takeoffs, emergency landings, heart-attack and in-flight medical situations, smoke in the cabin, fires, depressurisation, on-board births and deaths, dangerous goods in the cabin as well as land and water landings including the preparation of passengers and the cabin, the emergency evacuation with evacuation slides and then the followup survival skills which include environments as open water, jungle, water, tropical and Arctic climates, along with a variety of emergency equipment. Flight attendants are now also given basic training on defense against terrorist attacks. (557 words)

NEW WORDS

purser	/ˈpəːsə/	n.	乘务长
checklist	/ˈtʃeklist/	n.	检查单
features	/ˈfiːtʃəz/	n.	装置
boarding	/ˈbɔːdiŋ/	n.	登机
turbulence	/ˈtəːbjuləns/	n.	颠簸气流
anticipate	/ænˈtisipeit/	v.	预期
lifevest	/ˈlaifvest/	n.	救生衣
flashlight	/ˈflæʃlait/	n.	手电筒
quantity	/ˈkwɒntiti/	n.	数量
unserviceable	/ˈʌnˈsəːvisəbl/	adj.	不工作的

rectify	/ˈrektifai/	v.	纠正
monitor	/ˈmɒnitə/	v.	监控
precaution	/priˈkɔːʃən/	n.	预防
disarm	/disˈɑːm/	v.	解除预位
fuelling	/ˈfjuːəliŋ/	n.	加油
load	/ləʊd/	n.	负载；负荷
		v.	装载
baggage	/ˈbægidʒ/	n.	行李
demonstration	/ˌdemənˈstreiʃən/	n.	演示
stow	/stəʊ/	v.	收放
upright	/ˈʌpˈrait/	adj.	直立
armrest	/ˈɑːmrest/	n.	扶手
seatbelt	/ˈsiːtbelt/	n.	安全带
fasten	/ˈfɑːsn/	v.	系
lavatory	/ˈlævətəri/	n.	洗手间
deactivate	/ˌdiːˈæktiveit/	v.	断开
garbage	/ˈgɑːbidʒ/	n.	垃圾
mechanical	/miˈkænikəl/	adj.	机械的
station	/ˈsteiʃən/	n.	站位；电台
exit	/ˈeksit/	n.	出口
disembark	/ˌdisimˈbɑːk/	v.	下飞机
escort	/ˈeskɔːt/	v.	护送
drunken	/ˈdrʌkən/	adj.	醉酒的
depressurisation	/ˌdipreʃəraiˈzeiʃən/	n.	释压
evacuation	/iˌvækjuˈeiʃən/	n.	撤离
tropical	/ˈtrɒpikəl/	adj.	热带的
capacity	/kəˈpæsiti/	n.	容量

EXPRESSIONS

safety check　安全检查	on board　飞机上
fire fighting　消防	dangerous goods　危险品
tray table　桌板	smoke detector　烟雾探测器
call light　呼叫灯	crosscheck　交叉检查
severe turbulence　严重颠簸气流	rejected takeoff　中断起飞
evacuation slide　撤离滑梯	arctic climate　极地气候
safety features　安全装置	terrorist attack　恐怖袭击

Lesson 2

Cockpit and CRM

Preview

> **The airplane, a giant whole of complex systems controlled from the cockpit where instruments, controls and other cockpit resources are located, should be well managed by pilots with the help of cockpit resources management.**

Text A

Warming-up Activities

Picture Description

Please describe the following picture and be prepared to answer some questions.

Relevant Questions

1. Where do pilots work in an airplane?
2. In which flight phase of flight is the aircraft? Why?
3. Have you learned anything about crew resource management (CRM)?
4. Do you like the cockpit of Cessna 172 or that of Airbus 320? Why?

Cockpit

A cockpit or flight deck is the area, usually near the front of an aircraft, from which a pilot controls the aircraft. Most modern cockpits are enclosed, except on some small aircraft, and cockpits on large airliners are also separated from the cabin. The cockpit of an aircraft contains flight instruments on an instrument panel, and the controls which enable the pilot to fly the aircraft. In most airliners, a door separates the cockpit from the passenger compartment.

Cockpit windows may be equipped with a sun shield. Most cockpits have windows which can be opened when the aircraft is on the ground. Nearly all glass windows in large aircraft have an anti-reflective coating, and an internal heating element to melt ice. In most cockpits the pilot's control column is located centrally, although in some military fast jets and in some commercial airliners the pilot uses a side-stick.

Human factors are important in the design of modern cockpits. The layout and function of cockpit displays and controls are designed to increase a pilot's situation awareness without

causing information overload.

In the design of the cockpit in a military fast jet, the traditional "knobs and dials" associated with the cockpit are mainly absent. Instrument panels are now almost wholly replaced by electronic displays which are often re-configurable to save space. While some hard-wired dedicated switches must still be used for reasons of integrity and safety, many traditional controls are replaced by multi-function controls. Controls are incorporated onto the stick and throttle to enable the pilot to maintain a head-up and eyes-out position. New advances in auditory displays even allow for Direct Voice Output of aircraft status information and for the spatial localization of warning sounds for improved monitoring of aircraft systems.

The layout of control panels in modern airliners has become largely similar across the industry. The majority of the systems-related controls are usually located on an overhead panel. Radios are generally placed on a panel between the pilot's seats known as the pedestal. Automatic flight controls such as the autopilot are usually placed just below the windscreen and above the main instrument panel on the glareshield.

The primary component of the glass cockpit is the Electronic Flight Instrument System (EFIS), which displays all information regarding the aircraft's situation, position and progress. Comprising left-and-right-side primary flight display (PFD) and multi-function display (MFD) screens, EFIS primarily covers horizontal and vertical position, but also indicates time and speed. On the flight deck, the display units are the most obvious parts of an EFIS system, and are the features which give rise to the name "glass cockpit". The display unit taking the place of the ADI is  called the primary flight display. If a separate display replaces the HSI, it is called the navigation display (ND). The PFD displays all information critical to flight, including calibrated airspeed, altitude, heading, attitude, vertical speed and yaw. The PFD is designed to improve a pilot's situational awareness by integrating this information into a single display instead of six different analog instruments, reducing the amount of time necessary to monitor the instruments. PFDs also increase situational awareness by alerting the aircrew to unusual or potentially hazardous conditions by changing the color or shape of the display or by providing audio alerts.

The second part of the glass cockpit, comprising over-and-under center display screens, shows the aircraft's systems conditions and engines performance. This is variously called EICAS (Engine Indications and Crew Alerting System) or ECAM (Electronic Centralised Aircraft Monitor), the former being the Boeing term and the latter Airbus' acronym. All this information is graphically presented in a "need-to-know" basis, however, the pilot may query the system for further details of interest.

In the cockpit, the EFIS has some unique advantages. EFIS offers versatility by avoiding some of the physical limitations of traditional instruments. Thus, the same display which shows a course deviation indicator (CDI), can be switched to show the planned track provided by an area navigation or flight management system. If desired, the weather radar picture can then be superimposed on the displayed route. Another advantage is that the flexibility afforded by software modifications minimizes costs when new aircraft equipment and new regulations are introduced. The EFIS system can be updated with new software to extend its capabilities. Such updates introduced in the 1990s included enhanced GPWS, and TCAS. Finally, a degree of redundancy is available even with the simple two-screen EFIS installation. Should the PFD fail, transfer switching repositions its vital information to the screen normally occupied by the navigation display. (756 words)

NEW WORDS

enclose	/ in'kləuz/	v.	密闭
deck	/dek/	n.	甲板
instrument	/'instrumənt/	n.	仪表
panel	/'pænəl/	n.	面板
compartment	/kəm'pɑ:tmənt/	n.	舱
coating	/'kəutiŋ/	n.	涂层
overload	/ˌəuvə'ləud/	n.	过载
knob	/nɒb/	n.	旋钮
dial	/'daiəl/	n.	仪表盘
replace	/ri'pleis/	v.	取代；更换
electronic	/ˌelik'trɒnik/	adj.	电子的
switch	/switʃ/	n. v.	开关；转换
integrity	/in'tegriti/	n.	统一
throttle	/'θrɒtl/	n.	油门
augment	/ɔ:g'ment/	v.	增加，增益
auditory	/'ɔ:ditəri/	adj.	声音的，听觉的
status	/'steitəs/	n.	状态
pedestal	/'pedistl/	n.	操纵台
autopilot	/'ɔ:təuˌpailət/	n.	自动驾驶仪
windscreen	/'windskri:n/	n.	风挡
glareshield	/'gleəˌʃi:ld/	n.	遮光板
component	/kəm'pəunənt/	n.	部件，组件

progress	/ˈprəʊgres/	*n.*	进程
horizontal	/ˌhɒriˈzɒntl/	*adj.*	水平的
vertical	/ˈvəːtikl/	*adj.*	垂直的
navigation	/ˌnævɪˈgeiʃən/	*n.*	导航，领航
calibrate	/ˈkælibreit/	*v.*	修正，校准
airspeed	/ˈeəˈspiːd/	*n.*	空速
altitude	/ˈæltitjuːd/	*n.*	高度
yaw	/jɔː/	*n.*	偏航
analog	/ˈænəlɒg/	*adj.*	模拟式的
alert	/əˈləːt/	*n. & v.*	警告，警觉
hazardous	/ˈhæzədəs/	*adj.*	危险的
descent	/diˈsent/	*n.*	下降
audio	/ˈɔːdiəʊ/	*n.*	音频
performance	/pəˈfɔːməns/	*n.*	性能
acronym	/ˈækrənim/	*n.*	缩略词
query	/ˈkwiəeri/	*v.*	询问
versatility	/ˌvəːsəˈtiliti/	*n.*	多功能，多能力
course	/kɔːs/	*n.*	航道
deviation	/ˌdiːviˈeiʃən/	*n.*	偏离
track	/træk/	*n. & v*	航迹；跟踪
modification	/ˌmɒdifiˈkeiʃən/	*n.*	修改
update	/ʌpˈdeit/	*n. & v.*	更新
redundancy	/riˈdʌndənsi/	*n.*	余度，多余

EXPRESSIONS

flight deck　驾驶舱　　　　　　　　　　side-stick　侧杆(空客)

control column　操纵杆　　　　　　　　flight management system　飞行管理系统

flight instrument　飞行仪表　　　　　　course deviation indicator　航道偏离指示器

sun shield　遮光板

NOTES

ADI	Attitude Direction Indicator	姿态指引仪
HSI	Horizontal Status Indicator	水平状态指示器
GPWS	Ground Proximity Warning System	近地警告系统
TCAS	Traffic Collision Avoidance System	防撞系统
PFD	Primary Flight Display	主飞行显示器

EXERCISES

Comprehension of Text A

I. Describe the given aviation terms from Text A in English.

1. flight deck
2. situational awareness
3. human factors
4. GPWS
5. EFIS
6. EICAS

II. Answer the following questions after you have read Text A.

1. What is contained in the cockpit according to the first paragraph?

2. How is the cockpit window protected from ice?

3. What is the purpose of the layout of cockpit displays and controls?

4. Why are cockpit controls incorporated onto the stick and throttle?

5. What is told in the passage about new advances in auditory displays?

6. Where are the systems-related controls located in the cockpit?

7. What is the primary component of the glass cockpit? What does it present?

8. In the passage, what is said about PFD?

9. What are the advantages of EFIS?

10. What systems were introduced to update the EFIS in 1990s?

Reading Aloud

III. Read the following paragraph aloud until you can say it in a natural way from your memory.

Stress Management

Any kind of emergency situation results from stress (压力), but there are also physical and mental stresses that a crew member might bring to a situation which may be difficult for others to detect (探测，发现). A crew member's over-all fitness to fly may be affected because of fatigue (疲劳), mental or emotional problems, to the extent that other crew members should be on the alert for any performance decline or incapacitation.

Skills related to stress management refer not only to one's ability to find the stress in others but primarily to anticipate, recognize and cope with one's own stress as well. This would include psychological stresses resulting from crew scheduling, anxiety over career and achievement stresses, from inter-personal problems with either the cabin crew or other flight crew member, as well as from the home and work interface(界面). It also comes from some family events such as the death of a spouse, divorce or marriage, all of which represent major life changes.

Stress problems can be reduced by encouraging open and frank communications between operational management and flight crew members, and by regarding stress as part of "fitness to fly". Flight crew should be encouraged to understand stress problems and to attend CRM training. (207 words)

Vocabulary Practice

IV. Complete the following short passage by filling in the blanks with the words given in the box.

> accurate   decisions   emergency   tasks   human
> decision-making   impact   status   aviation   lacking

Situation Awareness

Situation awareness, or SA, is the perception of environmental elements within a volume（数量，体积）of time and space, the understanding of their meaning, and the projection（预示）of their 1_____ in the near future. It is also a field of study concerned with perception of the environment critical to decision-makers in complex, dynamic (动态的) areas from 2_____, air traffic control, power plant operations, military command and control–to more ordinary but nevertheless complex 3_____, such as driving an automobile or motorcycle.

Situation awareness (SA) involves being aware of what is happening around you to understand how information, events, and your own actions will 4_____ your goals and objectives(目的), both now and in the near future. 5_____ SA or having inadequate SA has been identified as one of the primary factors in accidents attributed to 6_____ error. Thus, SA is especially important in work domains where the information flow can be quite high and poor 7_____ may lead to serious consequences, such as piloting an airplane, functioning as a soldier, or treating critically ill or injured patients.

Having complete, 8_____ and up-to-the-minute SA is essential where technological and situational complexity (复杂) on the human decision-maker is a concern. SA has been recognized as a critical, yet often elusive (难以理解的), foundation for successful 9_____ across a broad range of complex and dynamic systems, including aviation and air traffic control, 10_____ response and military command and control operations, and offshore oil

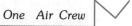

and nuclear power plant management. (236 words)

Reading for More

V. Read the following passage and answer the given questions briefly.

Standard Operating Procedures (SOPs)

Standard Operating Procedures (SOPs) are designed to enhance safety, to assist the flight crews in managing risks and ensuring consistency in the cockpit. Consider them guidelines as to who-does-what-and-when. At all times, these procedures should encourage effective communication and teamwork in the cockpit. SOPs can be either general in nature or aircraft specific. Although aviation companies have a certain amount of latitude (纬度;自由) when creating their SOPs, under no circumstances should they violate (违背) aviation regulations or the procedures outlined in the aircraft flight manuals (手册).

Included in the SOPs should be a section on normal procedures, and it is to be considered an enhancement of the aircraft flight manual. These normal procedures include all aspects of day-to-day flight, including the start-up and normal procedures checklist, the take-off briefing, go-around (复飞) procedures, IFR approaches, etc.

A section of the SOPs should be dedicated to emergency procedures, with the expanded version of the emergency procedures checklists. Again, this should compliment the aircraft flight manual's emergency section and streamline procedures to remove any ambiguity (模糊 不清).

Another section can include aircraft landing and take off distances charts, a JBI chart, or any other references that the flight crews may require.

SOPs should be reviewed periodically and amended to maintain their relevance in changing times or aircraft fleet(机队). Care should be taken not to include non-applicable items, as personnel may then tend to view the whole package as being irrelevant. SOPs should be written in simple terminology(术语), leaving no room for subjective interpretation.

There are definite safety benefits from the use of Standard Operating Procedures but they must first be adopted by the flight crews. Company check pilots should monitor for crew adherence to the SOPs. Finally, there is no substitute for good judgment, and decisions made in the cockpit should be supported by management. (296 words)

Question 1   Why are there SOPs in the cockpit?

Question 2   What is included in the SOPs?

Question 3   What should be done to ensure the proper adherence to SOPs?

VI. In this section, there are two passages. After you have read the two passages, there are five questions followed by four choices marked A, B, C and D. Decide which one is the most appropriate answer.

Passage One

Awareness of Maintaining Airplane Control

Once the pilot recognizes and accepts the situation, he or she must understand that the only way to control the airplane safely is by using and trusting the flight instruments. Attempts to control the airplane partially by reference to flight instruments while searching outside the cockpit for visual confirmation of the information provided by those instruments will result in inadequate airplane control. This may be followed by spatial disorientation (空间失定向性；迷航) and complete control loss.

The most important point to be stressed is that the pilot must not panic. The task at hand may seem overwhelming, and the situation may be compounded by extreme apprehension(忧虑). The pilot therefore must make a conscious effort to relax.

The pilot must understand the most important concern—in fact the only concern at this point—is to keep the wings level. An uncontrolled turn or bank usually leads to difficulty in achieving the objectives of any desired flight condition. The pilot will find that good bank control has the effect of making pitch control much easier.

The pilot should remember that a person cannot feel control pressures with a tight grip(紧紧握住) on the controls. Relaxing and learning to **"control with the eyes and the brain"** instead of only the muscles, usually take considerable conscious effort.

The pilot must believe what the flight instruments show about the airplane's attitude regardless of what the natural senses tell. The vestibular sense (motion sensing by the inner ear) can and will confuse the pilot. Because of inertia(惯性), the sensory areas of the inner ear cannot detect slight changes in airplane attitude, nor can they accurately sense attitude changes which occur at a uniform rate over a period of time. On the other hand, false sensations(错觉) are often generated, leading the pilot to believe the attitude of the airplane has changed when, in fact, it has not. These false sensations result in the pilot experiencing spatial disorientation. (316 words)

  1. Maintaining safe flying is only achieved by means of _____.
   A. outside reference   B. flight instruments
   C. visual confirmation   D. spatial orientation
  2. Faced with demanding tasks, the pilots are best advised not to _____.

A. relax   B. attempt   C. apprehend   D. panic

3. In complex situation the pilots only need to _____.

    A. keep wings level   B. make necessary turns

    C. ease control pressure   D. control the effect of pitch

4. "Control with the eyes and the brain" (Line 2, Para. 4) means that pilots should pay more attention to _____.

    A. manual control pressure   B. flight rules

    C. instrument indication   D. natural senses

5. Spatial disorientation often results from _____.

    A. pilot's confusion   B. false sensations

    C. attitude changes   D. instrument errors

Passage Two

Crew Resource Management (CRM)

CRM training covers a wide range of knowledge, skills and attitudes including communications, situational awareness, problem solving, decision making, and teamwork, together with all the attendant sub-disciplines to which each of these areas relate. CRM can be defined as a management system which makes optimum use of all available resources – equipment, procedures and people–to promote safety and enhance the efficiency of flight operations.

CRM is concerned not so much with the technical knowledge and skills required to fly and operate an aircraft but rather with the cognitive and interpersonal skills needed to manage the flight within an organized aviation system. In this context, cognitive (认知的) skills are defined as the mental processes used for gaining and maintaining situational awareness, for solving problems and for making decisions. Interpersonal skills are regarded as communications and a range of behavioral activities associated with teamwork. In aviation, as in other walks of life, these skill areas often overlap (重叠) with each other, and they also overlap with the required technical skills. Furthermore, they are not confined to multi-crew aircraft, but also relate to single pilot operations, which invariably need to interface with other aircraft and with various ground support agencies in order to complete their missions successfully.

CRM training for crew has been introduced and developed by aviation organizations including major airlines and military aviation worldwide. CRM training is now a compulsory requirement for commercial pilots working under most regulatory bodies worldwide, including the FAA (US) and JAA (Europe). Following the lead of the commercial airline industry, the US Department of Defense began formally training its air crews in CRM in the early 1990s. Presently, the US Air Force requires all air crew members to receive annual CRM

training, in an effort to reduce human-error caused mishaps (不幸). (294 words)

6. CRM is designed to _____.

   A. solve problems and ensure communication

   B. promote safety and enhance efficiency of flight operation

   C. include a wide range of resources

   D. incorporate the attendant sub-disciplines

7. According to the passage, what is more important in CRM?

   A. Solving problems and making decisions.

   B. Flying and operating an aircraft.

   C. The technical knowledge and skills.

   D. The cognitive and interpersonal skills.

8. In CRM, interpersonal skills is closely related to _____.

   A. situational awareness   B. knowledge and skills

   C. teamwork   D. mental processes

9. In aviation, interpersonal skills are useful to every pilot because he _____.

   A. has to interface with other aircraft and support agencies

   B. needs to learn from all walks of life

   C. is confined to the cockpit environment

   D. may have the chance of single operation

10. It can be concluded from the passage that CRM is _____.

   A. introductory   B. developed

   C. military   D. compulsory

VII. Read the following passage and retell it in your own words.

The A320 Family pioneered the new-generation Airbus flight deck, which introduced full **fly-by-wire**[1] controls, an efficient instrument panel and the **side-sticks**[2] that have become the preferred means of flight control for pilots.   [1] 电传飞行   [2] 侧杆

The A320 Family's advanced cockpit is virtually identical to that of the A330 and A340 wide-body aircraft, and it will retain a high degree of similarity with the new 555-seat A380. A single type rating enables pilots to fly the A318, A319, A320 and A321, while **transition**[3] to the larger A330, A340 and A380 is facilitated by the **commonality**[4] that exists between all Airbus fly-by-wire aircraft.   [3] 改装；过渡   [4] 通用性

Six interchangeable **liquid crystal displays**[5] (LCDs) on the main instrument panel replace the multitude of **dials**[6] and gauges on   [5] 液晶显示   [6] 仪表盘

older-generation aircraft, making it easy to assimilate information and ensuring a low **workload**[7] for the two-member crew. The fly-by-wire controls allow the pilot freedom within the aircraft's operating **envelope**[8], which protects against **overspeeding**[9], stalling and maintaining operation well within the aircraft's **structural**[10] limits. Side-stick controllers eliminate the **bulky**[11] centre-mounted control columns on other aircraft, opening an unobstructed view of the instrument panel and providing for the best use of fly-by-wire flight controls.

[7] 工作负荷
[8] 包线
[9] 超速
[10] 结构的
[11] 体积大的

In the A320, ECAM (Electronic Centralised Aircraft Monitor) is a system that monitors aircraft functions and **relays**[12] them to the pilots. It also produces messages detailing failures and in certain cases, lists procedures to undertake to correct the problem.

[12] 继电器；
传递；转发

ECAM is similar to another system, known as Engine Indicating and Crew Alerting System (EICAS), used by Boeing and others, which displays data concerning aircraft systems and also failures. Airbus developed ECAM, such that it not only provided the **features**[13] of EICAS, but also displayed corrective action to be taken by the pilot, as well as system limitations after the failures. Using a color coded scheme, the pilots can instantly assess the situation and decide on the actions to be taken. ECAM was first introduced in the Airbus A320 and has been incorporated it into all Airbus aircraft produced since then. It was designed to ease pilot stress in abnormal and emergency situations, by designing a paperless cockpit in which all the procedures are instantly available.

[13] 装置

ECAM is actually a series of systems designed to display information to the pilots in a quick and effective manner. Sensors placed throughout the aircraft, monitoring key **parameters**[14], feed their data into two SDACs (System Data Acquisition Concentrator) which in turn process the data and feed it to two FWCs (Flight Warning Computers.) The FWCs check for discrepancies in the data and then display the data on the ECAM displays through the three Display Management Computers (DMCs). In the event of a fault the FWCs generate the appropriate warning messages and sounds. More vital systems are routed directly through the FWCs such that failures in them can still be detected even with the loss of both SDACs. The whole system can continue to operate even with a failure of one SDAC and one FWC.(456 words)

[14] 参数

Text B

Passage Writing

VIII. Read the following passage carefully and write a passage under the given title with no less than 200 words.

Factors Affecting Safe Flying

Risk Assessment（风险评估）

Every aspect of life involves some element of risk, regardless of whether you drive a car, ride a motorcycle or fly an airplane. You must learn to cope with the risks associated with flying to ensure years of safe flying. The five subject areas – pilot, aircraft, environment, operation and situation–are also the five elements of risk. A close study of them is essential for student pilots and airline pilots.

As a pilot, your performance can be affected in many ways during a flight. Environment considerations, your physical conditions, and your emotional state can all influence your performance.

The risk element of the aircraft focuses on its equipment, condition, and suitability for the intended purpose of the flight. The best time to make this assessment is on the ground during preflight planning. However, it also needs to be done continuously during flight, since conditions can change at any time. For example, winds aloft may increase your anticipated flight time making available fuel a factor to be analyzed.

The environment is a far-reaching risk element which includes situations outside the aircraft that might limit, modify or affect the aircraft, pilot, and operational elements. Weather is a common environmental risk factor. Density altitude, runway length, obstacles, and related factors can also create environmental concerns.

In terms of the purpose of the flight operation, you must evaluate the interaction of you, your aircraft, and the environment. When determining the risks involved with beginning or continuing the flight as planned, be sure to consider all available concerns.

The combination of these first four risk elements leads into the fifth – the overall situation which you must evaluate continuously. Remember to become a safer pilot, you should increase your situational awareness.

Good aeronautical decision making requires a continuous assessment of whether to start a particular flight or to continue a flight as planned. You use the five risk elements to help you make a "go/no go" decision. Unless all five indicate "go", you should reevaluate your decision to make a flight. A good decision maker does not act hastily on feelings. With an accurate assessment of the risks associated with each of the five elements, you are best able to arrive at decisions that ensure a safe conclusion to a flight, even if it means not departing.

To assess risks effectively, you must be aware of risk raisers and the possibilities for risk accumulation, so you can determine the need to neutralize, or balance these factors. One way to become aware of this is to look at statistics to see what types of flight activities are most likely to result in accidents. The National Transportation Safety Board[1] (NTSB) has conducted studies of accident rates of the various types of flying in general aviation. They reveal that aerial application operations have the highest accident rate, followed by personal and business flying.

You should also be aware that the accident rate for single-engine airplanes is the highest for all general aviation airplane operations. More importantly, studies of the most common cause/factors of accidents in fixed wing aircraft show that most accidents were the result of an unsafe outcome of the pilot's decision-making process. In fact, 85% of all general aviation accidents can be attributed, at least in part, to pilot failure to maintain airspeed, misjudged distance, fuel exhaustion, inadequate preflight preparation and/or planning, selection of unsuitable terrain for landing, and inadequate visual scanning.

The NTSB also evaluated the phase of operation in which accidents occurred. The results indicate that the largest number occurred during landing, while 21.5% occurred during takeoff. Other statistics concern accident rates for pilots without instrument ratings flying single-engine airplanes at night in VFR conditions. (598 words)

NEW WORDS

risk	/risk/	*n.*	风险
		vt.	冒险……
motorcycle	/'məutə,saikl/	*n.*	摩托车
assessment	/ə'sesment/	*n.*	评价，评估
aspect	/'æspekt/	*n.*	方面

airline	/ˈeəlain/	n.	航空公司；航线
associate	/əˈsəuʃieit/	v.	联系（with）
ensure	/inˈʃuə/	v.	确保；保证
performance	/pəˈfɔ:məns/	n.	表现；性能
preflight	/priˈflait/	n.	飞行前
aloft	/əˈlɔft/	adj.	高空的
fuel	/fjuəl/	n.	燃油
density	/ˈdensiti/	n.	密度
obstacle	/ˈɒbstəkəl/	n.	障碍物
terrain	/ˈterein/	n.	地形
aeronautical	/ˌeərəˈnɔ:tikəl/	adj.	航空的
depart	/ˌdiˈpɑ:t/	v.	起飞，离场
interaction	/ˌintəˈækʃən/	n.	互动；交流
hastily	/ˈheistili/	adv.	匆忙地；草率地
accumulation	/əˌkju:mjuˈleiʃən/	n.	积累
activities	/ækˈtivitiz/	n.	活动 (pl.)
neutralize	/ˈnju:trəˌlaiz/	v.	中立
balance	/ˈbæləns/	n.	平衡
aviation	/ˌeiviˈeiʃən/	n.	航空
aerial	/ˈeəriəl/	adj.	航空的
airspeed	/ˈeəˌspi:d/	n.	空速
exhaustion	/igˈzɔ:stʃən/	n.	耗尽
landing	/ˈlændiŋ/	n.	着陆
scanning	/ˈskæniŋ/	n.	扫视
phase	/feiz/	n.	阶段
rating	/ˈreitiŋ/	n.	等级

EXPRESSIONS

cope with　处理；妥善处理

physical condition　身体条件

focus on　集中注意力于……

general aviation　通用航空

accident rate　事故率

fuel exhaustion　燃油耗尽

decision making　决策

safe flying　安全飞行

emotional state　情感状态

situational awareness　情景意识

business flying　商务飞行

preflight preparation　飞行前准备

instrument rating　仪表等级

ONLINE RESOURCES

National Transportation Safety Board (NTSB) (美国国家交通安全委员会) – The National Transportation Safety Board (NTSB) is an independent US Government investigative agency responsible for civil transportation accident investigation. In this role, the NTSB investigates and reports on aviation accidents and incidents, certain types of highway crashes, ship and marine accidents, pipeline incidents, and railroad accidents. When requested, the NTSB will assist the military with accident investigation. ***http://www.ntsb.gov/***

Unit Two
Aircraft Systems

Lesson 3

Aircraft Controls & Aerodynamics

Preview

Flight control surfaces utilize aerodynamic forces to lift the airplane off the ground and to support it in the air. Performance of airplane also depends upon correct opration of flight controls. Basic knowledge of fight controls and aircraft aerodynamics shall be discussed in this unit.

Text A

Warming-up Activities

Picture Description

Please describe the following picture and be prepared to answer some questions.

Relevant Questions

1. What courses have you learned about aircraft?
2. What parts does an airplane have?
3. Would you please introduce Cessna 172 briefly?
4. Which type of aircraft would you like to fly for your initial training? Why?

Airplane Flight Controls

The primary flight control system uses conventional control wheel, column and pedals linked mechanically to hydraulic power control units which command the primary flight control surfaces, ailerons, elevators and rudder. The flight controls are powered by redundant hydraulic sources: system A and system B. Either hydraulic system can operate all primary flight controls. The ailerons and elevators may be operated manually if required. The rudder may be operated by the standby hydraulic system if system A and system B's pressure is not available.

The secondary flight controls, high lift devices consisting of trailing edge (TE) flaps and leading edge (LE) flaps and slats (LE devices), are powered by hydraulic system B. In the event hydraulic system B fails, the TE flaps can be operated electrically. Under certain conditions the power transfer unit (PTU) automatically powers the LE devices. They can also be extended using standby hydraulic pressure.

The roll control surfaces consist of hydraulically powered ailerons and flight spoilers, which are controlled by rotating either control wheel.

The ailerons provide roll control around the airplane's longitudinal axis. The ailerons are positioned by the pilots' control wheels. The A and B FLT CONTROL switches control hydraulic shutoff valves. These valves can be used to isolate each aileron, as well as the elevators and rudder, from related hydraulic system pressure.

The Captain's control wheel is connected by cables to the aileron power control units (PCUs) through the aileron feel and centering unit. The First Officer's control wheel is connected by cables to the spoiler PCUs through the spoiler mixer. The two control wheels are connected by a cable drive system which allows actuation of both ailerons and spoilers by either control wheel. With total hydraulic power failure, the ailerons can be mechanically positioned by rotating the pilots' control wheels. Control forces are higher due to friction and aerodynamic loads.

If the ailerons or spoilers are jammed, force applied to the Captain's and the First Officer's control wheels will identify which system, ailerons or spoilers, is usable and which control wheel, Captain's or First Officer's, can provide roll control. If the aileron control system is jammed, force applied to the First Officer's control wheel provides roll control from the spoilers. The ailerons and the Captain's control wheel are inoperative. If the spoiler system is jammed, force applied to the Captain's control wheel provides roll control from the ailerons. The spoilers and the First Officer's control wheel are inoperative.

The pitch control surfaces consist of hydraulically powered elevators and an electrically powered stabilizer. The elevators are controlled by forward or aft movement of the control column. The stabilizer is controlled by autopilot trim or manual trim.

The elevators provide pitch control around the airplane's lateral axis. The elevators are positioned by the pilots' control columns. The A and B FLT CONTROL switches control hydraulic shutoff valves for the elevators. Cables connect the pilots' control columns to elevator power control units (PCUs) which are powered by hydraulic system A and B. The elevators are interconnected by a torque tube. With loss of hydraulic system A and B, the elevators can be mechanically positioned by forward or aft movement of the pilots' control columns. Control forces are higher due to friction and aerodynamic loads.

In the event of a control column jam, an override mechanism allows the control columns to be physically separated. Applying force against the jam will breakout either the Captain's or First Officer's control column. Whichever column moves freely after the breakout can provide adequate elevator control. Although total available elevator travel is significantly reduced, there is sufficient elevator travel available for landing flare. Column forces are higher and exceed those experienced during manual reversion. If the jam exists during the landing phase, higher forces are required to generate sufficient elevator control to flare for landing. Stabilizer trim is available to counteract the sustained control column force.

Yaw control is accomplished by a hydraulically powered rudder and a digital yaw damper system. The rudder is controlled by displacing the rudder pedals. The yaw damping functions are controlled through the stall management/yaw damper (SMYD) computers.

The rudder provides yaw control about the airplane's vertical axis. The A and B FLT CONTROL switches control hydraulic shutoff valves for the rudder and the standby rudder.

Each set of rudder pedals is mechanically connected by cables to the input levers of the main and standby rudder PCUs[2]. The main PCU consists of two independent input rods, two individual control valves, and two separate actuators: one for Hydraulic system A and one for Hydraulic system B. The standby rudder PCU is controlled by a separate input rod and control valve and powered by the standby hydraulic system. All three input rods have individual jam override mechanisms that allows input commands to continue to be transferred to the remaining free input rods if an input rod or downstream hardware is hindered or jammed.

The flaps and slats are high lift devices that increase wing lift and decrease stall speed during takeoff, low speed maneuvering and landing. LE devices consist of four flaps and eight slats: two flaps inboard and four slats outboard of each engine. Slats extend to form a sealed or slotted leading edge depending on the TE flap setting. The TE devices consist of double slotted flaps inboard and outboard of each engine.

TE flap positions 1-15 provide increased lift; positions 15-40 provide increased lift and drag. Flaps 15, 30 and 40 are normal landing flap positions. Flaps 15 is normally limited to airports where approach climb performance is a factor. Runway length and conditions must be taken into account when selecting a landing flap position. To prevent excessive structural loads from increased Mach at higher altitude, flap extension above 20,000 feet should not be attempted. (961 words)

NEW WORDS

pedal	/'pedəl/	n.	脚蹬，踏板
link	/link/	n. & v.	连接，联路
hydraulic	/hai'drɔːlik/	adj.	液压的
aileron	/'eilərɔn/	n.	副翼
elevator	/'eliveitə/	n.	升降舵
rudder	/'rʌdə/	n.	方向舵
standby	/'stændbai/	n.	备用的
flap	/flæp/	n.	襟翼
slat	/slæt/	n.	缝翼

spoiler	/'spɔilə/	*n.*	扰流板
torque	/tɔːk/	*n.*	扭力
tube	/tjuːb/	*n.*	筒，管
rotate	/rəu'teit/	*v.*	转动，抬轮
valve	/vælv/	*n.*	活门
cable	/keibl/	*n.*	缆索，电缆
actuation	/,æktʃu'eiʃən/	*n.*	启动，作动
load	/ləud/	*n.*	负荷，负载
stabilizer	/'steibilaizə/	*n.*	安定面
autopilot	/,ɔːtə'pailət/	*n.*	自动驾驶仪
trim	/trim/	*n. & v.*	配平
manual	/'mænjuəl/	*n.*	人工的
pitch	/pitʃ/	*n.*	俯仰
breakout	/'breikaut/	*n.*	断开
landing	/'lændiŋ/	*n.*	着陆
flare	/fleə/	*n.& v.*	拉平
friction	/'frikʃən/	*n.*	摩擦
reversion	/ri'vəːʃən/	*n.*	恢复
jam	/dʒæm/	*n.& v.*	卡阻
digital	/'didʒitəl/	*adj.*	数字式的
vertical	/'vəːtikl/	*adj.*	垂直的
rod	/rɒd/	*n.*	连杆
actuator	/'æktʃueitə/	*n.*	作动筒
override	/,əuvə'raid/	*v.*	超控
mechanism	/'mekənizm/	*n.*	机构，机械装置
transfer	/træns'fə:/	*v.*	转换
downstream	/'daun'striːm/	*n.*	下游
inboard	/'inbɔːd/	*adj.*	内侧
outboard	/'autbɔːd/	*adj.*	外侧
approach	/ə'prəutʃ/	*n.& v.*	进近
seal	/siːl/	*v.*	密封
extend	/iks'tend/	*v.*	伸出，放出
slot	/slɒt/	*n.*	缝
Mach	/mɑːk/	*n.*	马赫数

EXPRESSIONS

control wheel　驾驶盘　　　　　　high lift devices　增升装置

trailing edge　后缘	leading edge　前缘
longitudinal axis　纵轴	runway length　跑道长度
power transfer unit　动力转换装置	yaw damper　偏航阻尼器

NOTES

The hydraulic systems of aircraft vary. Those of smaller aircraft may be divided into System A and B. Typically B737 has left, center, and right hydraulic systems, while Airbus 320 has blue, green and yellow systems.

PCU (Power Control Unit): The rudder PCU consists of an input shaft/crank system, a dual concentric servo valve to control porting of the hydraulic fluid to the rudder actuator, and a yaw damper actuator.

EXERCISES

Comprehension of Text A

I. Describe the given aviation terms from Text A in English.

1. flight control surface　　　　2. secondary flight controls

3. pitch control　　　　　　　　4. roll control

5. structural load　　　　　　　6. yaw control

II. Answer the following questions after you have read Text A.

1. What devices are used to operate the primary flight control surfaces?

2. What components do secondary flight controls have?

3. What is the cockpit control used to operate pitch control?

4. How does control wheel operate roll control surfaces?

5. What are the pitch control surfaces?

6. What should the pilot do when the control column is jammed during landing?

7. What components does a rudder PCU include?

8. What is the basic function of jam override mechanisms?

9. What surfaces do high-lift devices include? What are their basic functions?

10. What factors may influence landing flap position?

Reading Aloud

III. Read the following paragraph aloud until you can say it in a natural way from your memory.

China Southern Airlines

China Southern Airlines is one of the largest companies in flight transportation, with its headquarters (总部) in Guangzhou, China. It has established 13 branches and 5 subsidiaries (子公司) in Xiamen, Guangxi, Guizhou and Zhuhai. It has several domestic operating offices and some foreign offices.

China Southern Airlines is a company with the largest number of airplanes, the most dense airline networks, and the largest annual passenger volumes (年旅客吞吐量). China Southern Airlines has many different types of airplanes dealing in transportation of passengers and cargos. They include Boeing aircraft like B777, 747, 757, 737 and Airbus airplanes, including A330, 321, 320, 319, 300. It has more than 600 airlines, going through more than 142 large and middle-size cities and covering various domestic and international airports. As far as I know, for many years, China Southern Airlines has had the largest number of passenger volume. (165words)

Vocabulary Practice

IV. Complete the following short passage by filling in the blanks with the words given in the box.

> circumstance   cause   excessive   varies   career   time
> maneuvers   training   stall   airplanes

Stall Awareness (失速意识)

14 CFR part 61 requires that a student pilot receive and log (记录) flight training in stalls and stall recoveries (改出) prior to solo flight. During this 1_____, the flight instructor should emphasize that the direct 2_____ of every stall is an excessive angle of attack (迎角). The student pilot should fully understand that there are some number of flight 3_____which may produce an increase in the wing's angle of attack, but the stall does not occur until the angle of attack becomes 4_____. This "critical" angle of attack (临界迎角) 5_____ from 16 to 20° depending on the airplane design.

The flight instructor must emphasize that low speed is not necessary to produce a

6_____ . The wing can be brought to an excessive angle of attack at any speed. High pitch attitude(俯仰姿态) is not an absolute indication of proximity(接近) to a stall. Some airplanes are capable of vertical flight with a corresponding low angle of attack. Most 7_____ are quite capable of stalling at a level or near level pitch attitude.

The key to stall awareness is the pilot's ability to visualize the wing's angle of attack in any particular 8_____ , and thereby be able to estimate his/her margin (余度) of safety above stall. This is a learned skill that must be acquired early in flight training and carried through the pilot's entire flying 9_____. The pilot must understand and appreciate factors such as airspeed, pitch attitude, load (负载) factor, relative wind, power setting, and aircraft configuration (外形) in order to develop a reasonably accurate mental picture of the wing's angle of attack at any particular 10_____ . It is essential to flight safety that a pilot takes into consideration this visualization of the wing's angle of attack prior to entering any flight maneuver. (287 words)

Reading for More

V. Read the following passage and answer the given questions briefly.

Thrust and Asymmetrical Thrust
(推力与不对称推力)

The basic means of controlling the aircraft is by making use of the position of the engine(s). If the engines are mounted (安装) under the centre of gravity (重心), as is the case in most passenger jets, then increasing the thrust will raise the nose, while decreasing the thrust will lower it. This control method may call for control inputs that go against the pilot's instinct: when the aircraft is in a dive (俯冲), adding thrust will raise the nose and vice versa (反之亦然).

Additionally, asymmetrical thrust may be used for directional control: if the left engine is idled(慢车) and power is increased on the right side, this will result in a yaw(偏航) to the left, and vice versa. If throttle (油门) settings allow the throttles to be shifted without affecting the total amount of power, then yaw control can be combined with pitch control. If the plane is yawing, then the wing on the outside of this yaw movement will go faster than the inner wing. This creates higher lift on the faster wing, resulting in a rolling (横滚) movement, which helps to make a turn.

Controlling speed is very difficult with engine control only, and will most likely result in a fast landing. A fast landing would be required anyway if the flaps can not be extended due to loss of hydraulics (液压). Only jet aircraft with an engine mounted on the vertical tail in addition to wing-mounted powerplants(动力装置) will be able to control the speed to a higher degree, as this engine is on the fuselage(机身) centerline and above the centre of gravity.

Aircraft that have two or four engines mounted on the sides of the empennage (尾翼) will only have limited benefit from asymmetrical thrust. Because this type of aircraft control is difficult for humans to achieve, some researchers have attempted to integrate this control ability into the computers of fly-by-wire aircraft. Early attempts to add the ability to real aircraft were not very successful, the software having been based on experiments conducted in flight simulators where jet engines are usually modeled as perfect devices with exactly the same thrust on each engine, a linear (线性) relationship between throttle setting and thrust, and instantaneous response to input. Later, computer models were updated to account for these factors, and aircraft have been successfully flown with this software installed. However, it remains a rarity (罕见) on commercial aircraft. (407 words)

Question 1　How does the position of engines relate to control input?

Question 2　What has got to be done if there is a loss of hydraulics?

Question 3　What efforts are made to take advantage of asymmetrical thrust?

VI. In this section, there are two passages. After you have read the two passages, there are five questions followed by four choices marked A, B, C and D. Decide which one is the most appropriate answer.

Passage One

Effects of Weight on Flight Performance

The takeoff/climb and landing performance of an airplane are determined on the basis of its maximum allowable takeoff and landing weights. A heavier gross weight (总重) will result in a longer takeoff run and shallower climb, and a faster touchdown speed and longer landing roll. Even a minor overload may make it impossible for the airplane to clear an obstacle that normally would not have been seriously considered during takeoffs under more favorable conditions.

The effects of overloading on performance are not limited to the immediate hazards involving takeoffs and landings. Overloading has an adverse (不利的) effect on all climb and cruise (巡航) performance which leads to overheating (超温) during climbs, added wear(磨损) on engine parts, increased fuel consumption (油耗), slower cruising speeds, and reduced range (航程).

The manufacturers of modern airplanes furnish weight and balance data with each airplane produced. Generally, this information may be found in the FAA approved Airplane Flight Manual or Pilot's Operating Handbook (AFM/POH). With the advancements in airplane design and construction in recent years has come the development of "easy to read charts" for determining weight and balance data. Increased performance and load-carrying capability of these airplanes require strict adherence to the operating limitations prescribed by the manufacturer. Deviations from the recommendations can result in structural damage or even complete failure of the airplane's structure. Even if an airplane is loaded well within the maximum weight limitations, it is imperative (强制的) that weight distribution (配重) be within the limits of center of gravity (重心). The preceding brief study of aerodynamics and load factors points out the reasons for this precaution. The following discussion is background information into some of the reasons why weight and balance conditions are important to the safe flight of an airplane.

The pilot is often completely unaware of the weight and balance limitations of the airplane being flown and of the reasons for these limitations. In some airplanes, it is not possible to fill all seats, baggage compartments, and fuel tanks, and still remain within approved weight or balance limits. As an example, in several popular four-place airplanes the fuel tanks may not be filled to capacity when four occupants and their baggage(行李) are carried. In a certain two-place airplane, no baggage may be carried in the compartment aft of the seats when spins are to be practiced. (381 words)

1. A heavier aircraft takes longer takeoff run because takeoff performance depends on _____.

    A. aircraft performance   B. takeoff weight

    C. empty weight   D. climb performance

2. To ensure obstacle clearance, it is advisable for pilots to _____.

    A. take off in good weather   B. perform longer takeoff roll

    C. avoid aircraft overload   D. apply more climb thrust

3. Increased fuel consumption is mentioned in the second paragraph to show _____.

    A. overloading has unfavorable effects on takeoff performance

    B. overloading has negative effects on climb and cruise performance

    C. in the phases of climb and cruise overload has less influences

    D. the trip range is shortened if fuel consumption is increased

4. To get increased performance, it is a must that airplanes obey _____.

    A. operating limitations   B. structural limitations

    C. weight limitations   D. CG limitations

5. From the passage we know that pilots should _____.

    A. practice spins in flight training

    B. fill the fuel tanks prior to each flight

C. take as few baggages as possible

D. know weight and balance performance

Passage Two

Thrust

Before the airplane begins to move, thrust must be exerted(施加). It continues to move and gain speed until thrust and drag (阻力) are equal. In order to maintain a constant airspeed, thrust and drag must remain equal, just as lift and weight must be equal to maintain a constant altitude. If in level flight, the engine power is reduced, the thrust is lessened, and the airplane slows down. As long as the thrust is less than the drag, the airplane continues to decelerate until its airspeed is insufficient to support it in the air.

Likewise, if the engine power is increased, thrust becomes greater than drag and the airspeed increases. As long as the thrust continues to be greater than the drag, the airplane continues to accelerate (加速). When drag equals thrust, the airplane flies at a constant airspeed.

Straight-and-level flight may be sustained at speeds from very slow to very fast. The pilot must coordinate angle of attack and thrust in all speed ranges if the airplane is to be held in level flight. Roughly, these ranges can be grouped in three categories: low-speed flight, cruising flight, and high-speed flight.

When the airspeed is low, the angle of attack must be relatively high to increase lift if the balance between lift and weight is to be maintained. If thrust decreases and airspeed decreases, lift becomes less than weight and the airplane will start to descend. To maintain level flight (平飞), the pilot can increase the angle of attack with an amount which will generate a lift force again equal to the weight of the airplane and while the airplane will be flying more slowly, it will still maintain level flight if the pilot has properly coordinated thrust and angle of attack.

Straight-and-level flight in the slow speed regime (范围) provides some interesting conditions relative to the equilibrium (等量) of forces, because with the airplane in a nose-high attitude, there is a vertical component (矢量) of thrust that helps support the airplane. For one thing, wing loading tends to be less than would be expected. Most pilots are aware that an airplane will stall, other conditions being equal, at a slower speed with the power on than with the power off. (Induced airflow over the wings from the propeller(螺旋桨) also contributes to this.) However, if analysis is restricted to the four forces as they are usually defined, one can say that in straight-and-level slow speed flight the thrust is equal to drag, and lift is equal to weight. (430 words)

6. In a constant speed flight, _____.

   A. thrust is equal to drag   B. weight is equal to lift

   C. engine power increases   D. engine power decreases

7. To maintain straight-and-level flight, it is very important to coordinate _____

   A. airspeed and altitude   B. angle of attack and thrust

   C. engine power and engine thrust   D. airplane weight and lift

8. In straight-and-level flight, changes in angle of attack serve to coordinate with changes in _____.

   A. weight   B. speed   C. thrust   D. drag

9. Why are most pilots aware that an airplane will stall, at a slower speed with the power on than with the power off?

   A. Because weight is equal to lift in straight-and-level flight.

   B. Because wing loading tends to be less than would be expected.

   C. Because the angle of attack is not coordinated with thrust.

   D. Because of induced airflow over the wings from the engines.

10. It can be inferred from the end of the passage that _____.

   A. induced airflow plays a more important role

   B. the propeller gives rise to induced airflow

   C. More than four forces act on the airplane in flight

   D. lift is properly adjusted to equalize weight at any time

VII. Read the following passage and retell it in your own words.

Lift

The pilot can control the lift. Any time the **control wheel**[1] is more fore or aft, the **angle of attack**[2] is changed. As angle of attack increases, lift increases (all other factors being equal). When the airplane reaches the maximum angle of attack, lift begins to **diminish**[3] rapidly. This is the **stalling angle of attack**[4]. The shape of the wing cannot be effective unless it continually keeps "attacking" new air. If an airplane is to keep flying, it must keep moving. Lift is proportional to the square of the airplane's **velocity**[5]. For example, an airplane traveling at 200 knots has four times the lift as the same airplane traveling at 100 knots, if the angle of attack and other factors remain constant.

Actually, the airplane could not continue to travel in **level flight**[6] at a constant altitude and maintain the same angle of attack if the velocity is

[1] 驾驶盘
[2] 迎角
[3] 减少
[4] 临界迎角
[5] 速度
[6] 平飞

44

increased. The lift would increase and the airplane would climb as a result of the increased lift force. Therefore, to maintain the lift and weight forces in balance, and to keep the airplane **"straight and level**[7]**"** (not accelerating[8] upward) in a state of **equilibrium**[9], as velocity is increased, lift must be decreased. This is normally accomplished by reducing the angle of attack, i.e., lowering the nose. Conversely, as the airplane is slowed, the decreasing velocity requires increasing the angle of attack to maintain lift sufficient to maintain flight. There is, of course, a limit to how far the angle of attack can be increased, if a **stall**[10] is to be avoided.

[7] 平直

[8] 加速

[9] 平衡

[10] 失速

Therefore, it may be concluded that for every angle of attack there is a corresponding **indicated airspeed**[11] required to maintain altitude in steady, unaccelerated flight–all other factors being constant. Since an airfoil will always stall at the same angle of attack, if increasing weight, lift must also be increased, and the only method for doing so is by increased velocity if the angle of attack is held constant just short of the "critical" or stalling angle of attack.

[11] 指示空速

Lift and **drag**[12] also vary directly with the density of the air. Density is affected by several factors: pressure, temperature, and **humidity**[13]. Remember, at an altitude of 18,000 feet, the density of the air has one-half the density of air at sea level. Therefore, in order to maintain its lift at a higher altitude, an airplane must fly at a greater true airspeed for any given angle of attack.

[12] 阻力

[13] 湿度

Furthermore, warm air is less dense than cool air, and **moist**[14] air is less dense than dry air. Thus, on a hot humid day, an airplane must be flown at a greater true airspeed for any given angle of attack than on a cool, dry day. If the density factor is decreased and the total lift must equal the total weight to remain in flight, it follows that one of the other factors must be increased. The factors usually increased are the airspeed or the angle of attack, because these factors can be controlled directly by the pilot. (505 words)

[14] 潮湿的

Text B

Passage Writing

VIII. Read the following passage carefully and write a passage under the given title with no less than 200 words.

Avoid Dangers from Wake Turbulence

_____

_____

Wake Turbulence

All aircraft produce wake turbulence, more correctly called wingtip vortices. Wake vortices are formed any time an aerofoil is producing lift. Lift is generated by the creation of a pressure differential over the wing surfaces. The lowest pressure occurs over the upper surface of the wing, and the highest pressure is formed under the wing. Air will always want to move towards the area of lower pressure. This causes it to move outwards under the wing towards the wingtip and curl up and over the upper surface of the wing. This starts the wake vortex.

The same pressure differential also causes air to move inwards over the wing. Small trailing edge vortices, formed by outward and inward moving streams of air meeting at the trailing edge, move outwards to the wingtip and join the large wingtip vortex. Swirling air masses trail downstream of the wingtips. Viewed from behind, the left vortex rotates clockwise and the right vortex rotates counter clockwise.

The greatest hazard from wake turbulence is induced roll and yaw. This is especially dangerous during takeoff and landing when there is little altitude for recovery. Aircraft with short wingspans are most affected by wake turbulence.

The effect of wake turbulence on an aircraft depends on many factors, including the weight and the wingspan of the following aircraft and relative positions of the following aircraft and wake vortices. In its mildest form, you may only experience a slight rocking of the wings, similar to flying through mechanical turbulence. In its most severe form, a complete loss of control of the aircraft may occur. The potential to recover from severe forms of wake turbulence will depend on altitude, maneuverability and power of your aircraft.

In general, you can expect induced roll and yaw. Small aircraft following larger aircraft may often be displaced more than 30 degrees in roll. Depending on the location of the trailing aircraft relative to the wake vortices, it is most common to be rolled in both directions.

The most dangerous situation is for a small aircraft to fly directly into the wake of a larger aircraft. This usually occurs while flying beneath the flight path of the larger aircraft. Flight tests conducted in this situation have shown that it is not uncommon for severe rolling motions to occur with loss of control. In other instances, if the aircraft is flown between the vortices, high roll rates can coincide with very high sink rates in excess of 1000 feet per minute. Depending on the altitude, the outcome could be tragic.

Flight tests conducted by pilots attempting to fly into the vortex at a slightly skewed angle resulted in a combination of pitching and rolling, which typically deflects the aircraft away from the wake. Research shows the greatest potential for a wake turbulence incident occurs when a light aircraft is turning from base to final behind a heavy aircraft flying a straight-in approach. The light aircraft crosses the wake vortices at right angles, resulting in short-lived pitching motions that can result in structural damage to the aircraft from a sudden increase in load factors.

The initial intensity of the wake vortices is determined by the weight, speed, configuration, wingspan and angle of attack of the aircraft. The most important variables in determining the intensity of the vortex beyond a distance of 10 to 15 wingspans from the aircraft are atmospheric stability, wind strength and direction, ground effect, and mechanical turbulence.

The strongest vortices are produced by heavy aircraft flying slowly in a clean configuration at high angles of attack. Considerable wake vortices can also be generated by maneuvering aircraft, for example, during aerobatics.

Aircraft with smaller wingspans generate more intense wake vortices than aircraft with equivalent weights and longer wingspans. The Boeing 757, for example, has a relatively short wing and large power plant for the weight of the aircraft. The wake turbulence that is produced by the 757 is equivalent to that of a much heavier aircraft.

Wake vortices near the ground are most persistent in light wind conditions (3 to 10 knots) in stable atmospheric conditions. Light crosswinds may cause the vortices to drift. A 3 to 5 knot crosswind will tend to keep the upwind vortex in the runway area and may cause the downwind vortex to drift toward another runway. Atmospheric turbulence generally causes them to break up more rapidly. (728 words)

NEW WORDS

wake	/weik/	n.	尾流
wingtip	/ˈwiŋtip/	n.	翼尖
vortice	/ˈvɔːtis/	n.	涡系
aerofoil	/ˈeərəfɔil/	n.	翼面
differential	/ˌdifəˈrenʃəl/	n.	差
curl	/kəːl/	v.	卷曲
swirl	/swəːl/	v.	旋转
inward	/ˈinwəd/	adj.	内向的
outward	/ˈautwəd/	adj.	外向的
maneuverability	/məˌnuvərəˈbiliti/	n.	机动性
hazard	/ˈhæzəd/	n.	危险
wingspan	/ˈwiŋspæn/	n.	翼展
recovery	/riˈkʌvəri/	n.	改出
motion	/ˈməuʃən/	n.	运动
pitch	/pitʃ/	n.	俯仰
base	/beis/	n.	四边；基地
final	/ˈfainəl/	n.	五边
		adj.	最后的
angle	/ˈæŋgl/	n.	角度
intensity	/inˈtensiti/	n.	强度
configuration	/ˌkɔnfigəˈreiʃən/	n.	构形，外形
variable	/ˈveəriəbəl/	n.	变量
aerobatics	/ˌeərəuˈbætiks/	n.	特技飞行
equivalent	/iˈkwivələnt/	adj.	相等的；等量的
drift	/drift/	v.& n.	偏流、偏移
upwind	/ʌpˈwind/	n.	逆风；一边
downwind	/ˌdaunˈwind/	n.	顺风；三边

EXPRESSIONS

pressure differential 压差	trailing edge 后缘
induced roll 诱导横滚	wingtip vortices 翼尖涡流
streams of air 气流	flight path 飞行轨迹
sink rate 下降率	right angle 直角
angle of attack 迎角	clean configuration 光洁外形
power plant 动力装置	

Lesson 4

Flight Control Systems & Problems

Preview

The main controls used to fly the airplane are the flight controls. In flight, loss of control is the biggest single cause of transport aircraft fatal accidents. Text A lists some primary causes for loss of control. Text B highlights the advantages of fly-by-wire control system over the mechanical and hydro-mechanical systems.

Text A

Warming-up Activities

Picture Description

Please describe the following picture and be prepared to answer some questions.

Relevant Questions

1. What are the flight control surfaces of an airplane?
2. Why are the wings of an aircraft important to flight?
3. How will you extend the landing gears both in normal and emergency situations?
4. If you have a flap setting problem during landing, how will the landing be possibly affected?

Loss of Control

Loss of control in flight has been one of the most significant causes of fatal aircraft accidents for many years. Loss of control usually occurs because the aircraft enters a flight regime which is outside its normal envelope, usually, but not always at a high rate, thereby introducing an element of surprise for the flight crew involved.

In flight, loss of control is the biggest single cause of transport aircraft fatal accidents. More attention to recovery from unusual attitudes for larger aircraft operating without a visual horizon reference is also needed, since a significant proportion of airborne loss of control accidents still occur when, if recognition of an abnormal aircraft attitude had been followed promptly by the optimum recovery action, a fatal outcome could still have been avoided. The "Loss of Control" issue for light aircraft is much more focused on marginal VFR weather and on the consequences of operating aircraft which are not certificated for flight in such conditions, both intentionally and unintentionally. Light aircraft pilots not adequately trained in instrument flight who end up in conditions which require it may also end up unable to retain aircraft control until a visual horizon can be re-acquired. The primary causes for loss of control are listed as follows.

A significant systems or systems control failure, which interferes with normal flight management and/or directly with aircraft control may lead to loss of control. This would include multiple engine failure, loss of correct function or control of a significant element of the flying controls, especially asymmetric spoilers/slats/flaps/thrust reversers, major electrical failure, and loss or malfunction of critical flight instrument displays.

The secondary result of structural failure and/or loss of power arising from a range of circumstances including midair collision, explosive decompression, fire on board or a wing fire, and contaminated or otherwise abnormal engine fuel feed may all lead to loss of control.

Crew incapacitation such that neither pilot is able to maintain control of the aircraft may lead to loss of control. This would include smoke and/or fumes in the flight deck and malfunction or incorrect control of the pressurization system. It might also occasionally include the consequences of deterioration in the physical or mental condition of just one of the pilots.

Loss of control may occur as a result of a flight management or control error or inappropriate intervention by or under the supervision of one or both of the pilots. This would include incorrect aircraft performance calculations, unintentional pilot mis-management of critical systems including engines, autopilot, fuel transfer, fuel exhaustion, pre-flight fuel loading, pilot disorientation under IMC or night VMC conditions and unintended operation outside the requirements of the AFM. It particularly also includes inappropriate or absent responses or inattention to otherwise relatively minor abnormalities which would not normally affect the safety of an aircraft.

Environmental factors external to the aircraft which interfere with normal use of engines, flight controls or critical flight instruments or lead to their capability being exceeded or cause other serious damage, can lead to loss of control. This would include ice accumulation on the airframe or sensors before takeoff or during flight, microburst/severe wind shear, severe wake vortex, severe air turbulence, the effects of ice entering or otherwise accreting within the engines and the effect on multiple engine function of passage through volcanic ash or an encounter with flocking birds resulting in bird strike. It could also include the effects of damage caused by runway surface debris of any origin which did not become apparent until after V1.

Loss of control can occur if the aircraft load is out of the allowable flight envelope or is mis-trimmed because the actual loading of the aircraft is not as documented. This would include in-flight load shift and fuel transfer effects as well as other pre-flight mis-loading scenarios including the loading of any items which should not be carried.

The effects of loss of control may include: discomfort or injury to the occupants prior to recovery to controlled flight, structural damage to, or total loss of, the aircraft, fatal or serious injury to occupants due to terrain impact and/or post impact fire. The effects of loss of control depend on the ability of the pilots to recover from the situation. This, in turn, depends on the nature of the upset causing loss of control, the experience and ability of the pilots and the height of the aircraft being adequate.

There are some effective measures to defend against loss of control. Multi crew pilot training stresses the need for an effective monitoring role for the "Pilot Not Flying" and any other members of the flight crew. Pilot training stresses the need to avoid distraction from the primary task of managing or flying the aircraft, especially when dealing with in-flight abnormal or emergency conditions. Pilot training and procedures ensure that the necessary responses to imminent loss of control alerts such as stall ident /warning, bank angle and negative wind shear are followed promptly and fully. In addition, continued training puts emphasis on VFR pilots planning and conducting their flight to stay in VMC. (846 words)

NEW WORDS

fatal	/ˈfeitəl/	*adj.*	致命的
accident	/ˈæksidənt/	*n.*	事故
regime	/ˈredʒim/	*n.*	状态，方式
envelope	/ˈenviləup/	*n.*	包线
rate	/reit/	*n.*	比率
visual	/ˈviʒuəl/	*adj.*	目视的
horizon	/həˈraizən/	*n.*	地平线
proportion	/prəˈpɔːʃən/	*n.*	比例
abnormal	/æbˈnɔːməl/	*adj.*	非正常的
marginal	/ˈmɑːdʒinəl/	*adj.*	边缘的
icing	/ˈaisiŋ/	*adj.*	结冰
interfere	/ˌintəˈfiə/	*v.*	干扰
multiple	/ˈmʌltipl/	*adj.*	多个，多重的
asymmetric	/ˌeisiˈmetrik/	*adj.*	不对称的
electrical	/iˈlektrikəl/	*adj.*	电气的
malfunction	/mælˈfʌŋkʃən/	*n.*	故障
collision	/kəˈliʒən/	*n.*	相撞，撞机
decompression	/ˌdiːkəmˈpreʃən/	*n.*	释压
contaminated	/kənˈtæmineitid/	*adj.*	污染
fume	/fjuːm/	*n.*	烟雾
pressurization	/ˌpreʃəraiˈzeiʃən/	*n.*	增压
deterioration	/diˌtiriˈreiʃən/	*n.*	恶化，变差

intervention	/ˌintəˈvenʃən/	n.	接入，干预
supervision	/ˌsjuːpəˈviʒən/	n.	监管，管理
airframe	/ˈeəˌfreim/	n.	机体
sensor	/ˈsensə/	n.	传感器
microburst	/ˈmaikrəˌbəːst/	n.	微下击气流
exhaustion	/iɡˈzɔːstʃən/	n.	耗尽
disorientation	/disˌɔːrienˈteiʃən/	n.	迷失方向；迷航
encounter	/inˈkauntə/	v.	遭遇
debris	/ˈdebriː/	n.	碎片
scenario	/siˈnɑːriəu/	n.	情景；场景
discomfort	/disˈkʌmfət/	n.	不舒适
occupant	/ˈɒkjupənt/	n.	乘员
impact	/ˈimpækt/	n.	撞击
upset	/ʌpˈset/	n.	失去安定性
imminent	/ˈiminənt/	adj.	立刻的
ident	/ˈaidənt/	n.	识别，识别符
bank	/bæŋk/	n.	坡度

EXPRESSIONS

flight envelope	飞行包线	bird strike	鸟击
performance calculations	性能计算	multi crew	多人机组
volcanic ash	火山灰	bank angle	坡度角

NOTES

VFR	Visual Flight Rules	目视飞行规则
VMC	Visual Meteorology Conditions	目视气象条件
IMC	Instrument Meteorology Conditions	仪表气象条件
AFM	Airplane Flight Manual	飞机飞行手册

EXERCISES

Comprehension of Text A

I. Describe the given aviation terms from Text A in English.

1. loss of control

2. normal envelope

3. VFR weather   4. crew incapacitation

5. fuel exhaustion   6. flight envelope

II. Answer the following questions after you have read Text A.

1. What is the usual reason for loss of control?

2. In what conditions will light aircraft suffer from loss of control?

3. What are the primary reasons for loss of control?

4. How can flight crew deal with incapacitation in the cockpit?

5. According to the passage, which systems can be mis-managed by pilots?

6. What kinds of environmental factors can lead to loss of control?

7. How can aircraft be properly loaded in order to avoid loss of control?

8. What are the possible effects of loss of control?

9. How can we defend against loss of control?

10. Of the reasons for loss of control, which one do you think is the most common?

Reading Aloud

III. Read the following paragraph aloud until you can say it in a natural way from your memory.

The effects of loss of control may include discomfort or injury to the occupants prior to recovery to controlled flight, structural damage to, or total loss of, the aircraft, fatal or serious injury to occupants due to terrain impact and/or post impact fire. The effects of loss of control depend on the ability of the pilots to recover from the situation. This, in turn, depends on the nature of the upset causing loss of control, the experience and ability of the pilots, and the height of the aircraft being adequate.

There are some effective measures to defend against loss of control. Multi crew pilot training stresses the need for an effective monitoring role for the "Pilot Not Flying" and any other members of the flight crew. Pilot training stresses the need to avoid distraction from the primary task of managing or flying the aircraft, especially when dealing with in-flight abnormal or emergency conditions. Pilot training and procedures ensure that the necessary responses to imminent loss of control alerts such as stall ident/warning, bank angle and negative wind shear are followed promptly and fully. In addition, continued training puts emphasis on VFR pilots planning and conducting their flight to stay in VMC. (202 words)

Vocabulary Practice

IV. Complete the following short passage by filling in the blanks with the words given in the box.

> rudder   flap   excessively   extended   direction
> wing   attitude   aileron   position   stall

Asymmetrical Flaps
（不对称襟翼）

An asymmetric "split" flap situation is one in which one flap deploys (放出) or retracts (收上) while the other remains in 1_____. The problem is indicated by a pronounced roll toward the 2_____ with the least flap deflection (偏转) when wing flaps are 3_____retracted.

The roll encountered in a split flap situation is countered with opposite 4_____. The yaw caused by the additional drag created by the extended flap will require substantial opposite 5_____ , resulting in a cross-control condition. Almost full aileron may be required to maintain a wings-level 6_____ , especially at the reduced airspeed necessary for approach and landing. The pilot therefore should not attempt to land with a crosswind from the side of the deployed 7_____, because the additional roll control required to counteract (抵制) the crosswind may not be available.

The pilot must be aware of the difference in 8_____ speeds between one wing and the other in a split flap situation. The wing with the retracted flap will stall considerably earlier than the wing with the deployed flap. This type of asymmetrical stall will result in an uncontrollable roll in the 9_____of the stalled (clean) wing. If altitude permits, a spin (螺旋) will result.

The approach to landing with a split flap condition should be flown at a higher than normal airspeed. The pilot should not risk an asymmetric stall (不对称失速) and subsequent loss of control by flaring 10_____. Rather, the airplane should be flown onto the runway so that the touchdown occurs at an airspeed consistent with a safe margin (安全余度) above flaps-up stall speed. (262 words)

Reading for More

V. Read the following passage and answer the given questions briefly.

Total Flap Failure

The inability to extend the wing flaps will necessitate a no-flap approach and landing. In light airplanes, a no-flap approach and landing is not particularly difficult or dangerous. However, there are certain factors which must be considered in the execution of this maneuver (机动). A no-flap landing requires substantially more runway than normal. The increase in required landing distance could be as much as 50 percent.

When flying in the traffic pattern (起落航线) with the wing flaps retracted, the airplane must be flown in a relatively nose-high attitude to maintain altitude, as compared to flight with flaps extended. Losing altitude can be more of a problem without the benefit of the drag normally provided by flaps. A wider, longer traffic pattern may be required in order to avoid the necessity of diving to lose altitude and consequently building up excessive airspeed.

On final approach, a nose-high attitude can make it difficult to see the runway. This situation, if not anticipated, can result in serious errors in judgment of height and distance. Approaching the runway in a relatively nose-high attitude can also cause the perception (感觉) that the airplane is close to a stall. This may cause the pilot to lower the nose abruptly and risk touching down on the nosewheel.

With the flaps retracted and the power reduced for landing, the airplane is slightly less stable in the pitch and roll axes. Without flaps, the airplane will tend to float (拉飘) considerably during roundout (拉平). The pilot should avoid the temptation to force the airplane onto the runway at an excessively high speed. Neither should the pilot flare (拉平) excessively, because without flaps this might cause the tail to strike the runway. (276 words)

Question 1 Which factors need to be considered in the event of flap-up landing?

Question 2 What is the proper attitude in traffic pattern with flaps retracted?

Question 3 On flap-up landing, what should a pilot do to avoid tail strike?

VI. In this section, there are two passages. After you have read the two passages, there are five questions followed by four choices marked A, B, C and D. Decide which one is the most appropriate answer.

Passage One

Loss of Elevator Control

In many airplanes, the elevator is controlled by two cables (钢索, 电缆): a "down" cable and an "up" cable. Normally, a break or disconnect in only one of these cables will not result in a total loss of elevator control. In most airplanes, a failed cable results in a partial loss of pitch control. In the failure of the "up" elevator cable, the control yoke (操纵杆) will move aft easily but produce no response. Forward yoke movement, however, beyond the neutral (中立) position produces a nose-down attitude. Conversely, a failure of the "down" elevator cable, forward movement of the control yoke produces no effect. The pilot will, however, have partial control of pitch attitude with aft movement. When experiencing a loss of up-elevator control, the pilot can retain pitch control by:

• Applying considerable nose-up trim.

• Pushing the control yoke forward to attain and maintain desired attitude.

• Increasing forward pressure to lower the nose and relaxing forward pressure to raise the nose.

• Releasing forward pressure to flare for landing.

When experiencing a loss of down-elevator control, the pilot can retain pitch control by:

• Applying considerable nosedown trim.

• Pulling the control yoke aft to attain and maintain attitude.

• Releasing back pressure to lower the nose and increasing back pressure to raise the nose.

• Increasing back pressure to flare for landing.

Trim mechanisms (配平机构) can be useful in the event of an in-flight primary control failure. For example, if the linkage between the cockpit and the elevator fails in flight, leaving the elevator free to weathervane in the wind, the trim tab can be used to raise or lower the elevator, within limits. The trim tabs (调整片) are not as effective as normal linkage (连杆) control in conditions such as low airspeed, but they do have some positive effect – usually enough to bring about a safe landing. If an elevator becomes jammed (卡阻), resulting in a total loss of elevator control movement, various combinations of power and flap extension offer a limited amount of pitch control. A successful landing under these conditions, however, is problematical. (349 words)

1. For most airplanes, a cable break may lead to _____.
   A. a total loss of elevator control   B. inoperative "down" cable
   C. partial loss of elevator control   D. inoperative "up" cable

2. In which condition will the pilot have partial control of pitch attitude?
   A. Forward yoke movement beyond the neutral position.
   B. Applying considerable nose-up trim.
   C. A failure of the "up" elevator cable.
   D. A failure of the "down" elevator cable.

3. With loss of down-elevator control, the pilot can have pitch control by_____.
   A. pushing the control yoke forward

B. applying considerable nosedown trim

C. applying considerable nose-up trim

D. increasing forward pressure

4. "weathervane" in the last paragraph most probably means _____.

    A. detect weather conditions   B. avoid adverse weather

    C. move freely in the air   D. forecast weather conditions

5. We can learn from this passage that the trim tab_____.

    A. has limited effects in assisting primary controls

    B. is as effective as any primary flight controls

    C. is effective only when there is loss of elevators

    D. can be jammed when combined with power and flaps

Passage Two

Stabilizer trim switches on each control wheel actuate the electric trim motor through the main electric stabilizer trim circuit when the airplane is flown manually. With the autopilot engaged, stabilizer trim is accomplished through the autopilot stabilizer trim circuit. The main electric and autopilot stabilizer trim has two speed modes: high speed with flaps extended, and low speed with flaps retracted. If the autopilot is engaged, actuating either pair of stabilizer trim switches automatically disengages the autopilot. The stabilizer trim wheels rotate whenever electric stabilizer trim is actuated. The STAB TRIM MAIN ELEC cutout switch and the STAB TRIM AUTOPILOT cutout switch, located on the control stand, are provided to allow the autopilot or main electric trim inputs to be disconnected from the stabilizer trim motor.

Control column actuated stabilizer trim cutout switches stop operation of the main electric and autopilot trim when the control column movement opposes trim direction. When the STAB TRIM override switch is positioned to OVERRIDE, electric trim can be used regardless of control column position. Manual stabilizer control is accomplished through cables which allow the pilot to position the stabilizer by rotating the stabilizer trim wheels. The stabilizer is held in position by two independent brake systems. Manual rotation of the trim wheels can be used to override autopilot or main electric trim. The effort required to manually rotate the stabilizer trim wheels may be higher under certain flight conditions. Grasping the stabilizer trim wheel will stop stabilizer motion.

In the event the stabilizer is trimmed to the end of the electrical trim limits, additional trim is available through the use of the manual trim wheels. If manual trim is used to position the stabilizer beyond the electrical trim limits, the stabilizer trim switches may be used to return the stabilizer to electrical trim limits.

Stabilizer Position Indication and Green Band Stabilizer position is displayed in units on two STAB TRIM indicators located inboard of each stabilizer trim wheel. The STAB TRIM indicators also display the TAKEOFF green band indication. (335 words)

6. In automatic flight, stabilizer trim is achieved by _____.

    A. electric motor in the electrical system

    B. trim switches on the control wheels

    C. manual trim wheel in the flight deck

    D. the autopilot stabilizer trim circuit

7. From the first paragraph it can be learned that _____.

    A. each pilot has his own stabilizer trim switch on the control wheel

    B. autopilot trim differs from electric trim in its two speed modes

    C. manual trim can trigger aural stabilizer trim warning in the cockpit

    D. the electric trim can be used even with autopilots engaged

8. If the control column is moved in opposite direction to the trim direction, it will lead to _____.

    A. disengagement of both autopilots   B. loss of electric and autopilot trim

    C. activating the stabilizer trim motor   D. ineffective manual stabilizer trim

9. When the stabilizer is beyond the electrical trim limit, _____ can be used to move the stabilizer.

    A. automatic trim   B. electric trim

    C. stabilizer trim   D. trim wheel

10. Which of the following is the most appropriate title for this passage?

    A. Automatic Trim.   B. Electric Trim.

    C. Stabilizer Trim.   D. Manual Trim.

VII. Read the following passage and retell it in your own words.

Basic Flight Control System

Some mechanical flight control systems use **servo tabs**[1] to [1] 伺服调整片 provide aerodynamic assistance. Servo tabs are small surfaces **hinged**[2] to the control surfaces. The flight control mechanisms move [2] 铰链 these tabs, aerodynamic forces in turn move, or assist the movement of the control surfaces reducing the amount of mechanical forces needed. This arrangement was used in early **piston**[3]-engined transport aircraft [3] 活塞 and in early jet transports.

The complexity and weight of mechanical flight control systems increase considerably with the size and performance of the aircraft. Hydraulic power overcomes these **limitations**[4]. With hydraulic flight control systems, aircraft size and performance are limited by economics rather than a pilot's strength. Initially only partially **boost**ed[5] systems were used in which the pilot could still feel some of the aerodynamic loads on the surfaces (**feedback**[6]).

    [4] 限制

    [5] 增压的
    [6] 反馈

A hydromechanical flight control system has two parts. The mechanical **circuit**[7] links the cockpit controls with the hydraulic circuits. Like the mechanical flight control system, it consists of rods, cables, pulleys, and sometimes **chains**[8]. The hydraulic circuit has hydraulic pumps, **reservoir**s[9], filters, pipes, valves, and **actuator**s[10]. The actuators are powered by the hydraulic pressure generated by the pumps in the hydraulic circuit. The actuators convert hydraulic pressure into control surface movements. The servo valves control the movement of the actuators.

    [7] 环路

    [8] 链条
    [9] 液压油箱
    [10] 作动筒

The pilot's movement of a control causes the mechanical circuit to open the matching servo valve in the hydraulic circuit. The hydraulic circuit powers the actuators which then move the control surfaces. As the actuator moves the servo valve is closed by a mechanical feedback linkage which stops movement of the control surface at the desired position.

With purely mechanical flight control systems, the aerodynamic forces on the control surfaces are transmitted through the mechanisms and are felt directly by the pilot. This gives feedback of airspeed and aids flight safety. With hydro-mechanical flight control systems, however, the load on the surfaces cannot be felt and there is a risk of **overstress**ing[11] the aircraft through excessive control surface movement. To overcome this problem, **artificial feel**[12] systems are used. A **stick shaker**[13] is a device fitted to provide artificial stall warning in some aircraft with hydraulically-powered control systems.(355 words)

    [11] 使……超过压力
    [12] 人工感力
    [13] 抖杆器

Text B

Passage Writing

VIII. Read the following passage carefully and write a passage under the given title with no less than 200 words.

Do You Prefer to Fly A320 or B737?

_____

_____

Fly-by-wire Control Systems

Mechanical and hydro-mechanical flight control systems are heavy and require careful routing of flight control cables through the aircraft using systems of pulleys, cranks, wires and, with hydraulically-assisted controls, hydraulic pipes. Both systems often require redundant backup to deal with failures, which again increases weight. Furthermore, both have limited ability to compensate for changing aerodynamic conditions. Dangerous characteristics such as stalling, spinning and pilot-induced oscillation (PIO), which depends mainly on the stability and structure of the aircraft concerned rather than the control system itself, can still occur with these systems.

By using electrical control circuits combined with computers, designers can save weight, improve reliability, and use the computers to mitigate the undesirable characteristics mentioned above. Advanced modern fly-by-wire systems are also used to control otherwise unstable fighter aircraft.

The words "Fly-by-Wire" (FBW) imply an electrically-signaled only control system. However, the term is generally used in the sense of computer-configured controls, where a computer system is interposed between the operator and the final control actuators or surfaces. This modifies the manual inputs of the pilot in accordance with control parameters. These are carefully developed and validated in order to produce maximum operational effect without compromising safety.

A FBW aircraft can be lighter than a similar design with conventional controls. Partly due to the lower overall weight of the system components and partly because the natural aerodynamic stability of the aircraft can be relaxed, slightly for a transport aircraft and more for a maneuverable fighter, which means that the stability surfaces that are part of the aircraft structure can therefore be made smaller. These include the vertical and horizontal stabilizers (fin and tailplane) that are (normally) at the rear of the fuselage. If these structures can be reduced in size, airframe weight is reduced. The advantages of FBW controls were first exploited by the military and then in the commercial airline market. The Airbus series of airliners used full-authority FBW controls beginning with their A320 series and Boeing followed with their 777 and later designs.

Electronic fly-by-wire systems can respond flexibly to changing aerodynamic conditions, by tailoring flight control surface movements so that aircraft response to control inputs is appropriate to flight conditions. Electronic systems require less maintenance, whereas mechanical and hydraulic systems require lubrication, tension adjustments, leak checks, fluid changes, etc. Furthermore, putting circuitry between pilot and aircraft can enhance safety; for example, the control system can try to prevent a stall, or it can stop the pilot from over stressing the airframe.

A fly-by-wire system actually replaces manual control of the aircraft with an electronic interface. The movements of flight controls are converted to electronic signals, and flight control computers determine how to move the actuators at each control surface to provide the expected response. The actuators are usually hydraulic, but electric actuators have been used.

While traditional mechanical or hydraulic control systems usually fail gradually, the loss of all flight control computers could immediately render the aircraft uncontrollable. For this reason, most fly-by-wire systems incorporate either redundant computers (triplex, quadruplex. etc.) to increase reliability, some kind of mechanical or hydraulic backup or a combination of both. A "mixed" control system such as the latter is not desirable and modern FBW aircraft normally avoid it by having more independent FBW channels, thereby reducing the possibility of overall failure to minuscule levels that are acceptable to the independent regulatory and safety authority responsible for aircraft design, testing and certification before operational service. (572 words)

NEW WORDS

hydro-mechanical	/ˌhaidrəˌmiˈkænikəl/	*adj.*	液压机械的
routing	/ˈruːtiŋ/	*n.*	管路
cable	/ˈkeibl/	*n.*	钢索；电缆

crank	/kræŋk/	*n.*	曲轴
wire	/waiə/	*n.*	电线
pipe	/paip/	*n.*	管道
backup	/ˈbækʌp/	*adj.*	备用的，辅助的
compensate	/ˈkɒmpenseit/	*v.*	补偿
aerodynamic	/ˌeərəudaiˈnæmik/	*adj.*	空气动力的
spin	/spin/	*n. & v.*	螺旋
oscillation	/ˌɒsiˈleiʃən/	*n.*	震荡
stability	/stəˈbiliti/	*n.*	稳定性
circuit	/ˈsəːkit/	*n.*	电路；环路
mitigate	/ˈmitiˌgeit/	*v.*	减小；减轻
interpose	/ˌintəˈpəuz/	*v.*	介于……之间
operator	/ˈɒpəˌreitə/	*n.*	操作人员
modify	/ˈmɒdifai/	*v.*	修改
manual	/ˈmænjuəl/	*adj.*	人工的
		n.	手册
parameter	/pəˈræmitə/	*n.*	参数
quadruplex	/ˈkwɒdrupleks/	*n.*	四倍
channel	/tʃænəl/	*n.*	频道；信道
negative	/ˈnegətiv/	*adj.*	负的；否定的
component	/kəmˈpəunənt/	*n.*	部件；矢量
stabilizer	/ˈsteibilaizə/	*n.*	安定面
fin	/fin/	*n.*	立尾
tailplane	/ˈteilplein/	*n.*	尾翼
fuselage	/ˈfjuːzəlɑːʒ/	*n.*	机身
tailor	/ˈteilə/	*v.*	调整
maintenance	/ˈmeintənəns/	*n.*	维护，维修
lubrication	/ˌluːbriˈkeiʃən/	*n.*	润滑
leak	/liːk/	*n.*	泄漏
fluid	/fluːid/	*n.*	液体
circuitry	/ˈsəːkitri/	*n.*	电路
convert	/kənˈvəːt/	*v.*	转换
triplex	/ˈtripleks/	*n.*	三倍
minuscule	/ˈminəskjul/	*adj.*	微小的
certification	/ˌsəːtifiˈkeiʃən/	*n.*	认证，证实

Lesson 5

Hydraulic Power & Landing Gears

Preview

Hydraulic power is one of the power sources which are widely used in modern airplanes to operate various airplane units and mechanisms; the landing gear is the assembly that supports the aircraft during landing or while it is resting or moving about on the ground. Text A shows us the primary components and functions of the hydraulic system. Text B mentions the controls in case of landing gear malfunction.

Text A

Warming-up Activities

Picture Description

Please describe the following picture and be prepared to answer some questions.

Relevant Questions

1. Does this airplane have tricycle landing gears or conventional gears?
2. What are the functions of landing gears on the ground?
3. What do you think is the most common reason for gear fires?
4. As a captain, how would you react to the situation in which the nose gear is jammed during final approach?

The Hydraulic System

Hydraulic power is one of the power sources which are widely used in modern airplanes to operate various airplane units and mechanisms. Hydraulic systems are not new to aviation. Early airplanes had hydraulic brake system. As the airplane became sophisticated, newer systems with hydraulic power were developed.

Although some airplane manufacturers make greater uses of hydraulic systems than others, the hydraulic system of the average modern airplane performs many functions. Among the units commonly operated by hydraulic systems are landing gear, wing flaps, slats, speed and wheel brakes, nose wheel steering and primary flight control surfaces.

A basic hydraulic system consists of the following components: a reservoir, an engine driven pump, a filter, a pressure regulator, an accumulator, a pressure gage, a relief valve, an emergency pump, two check valves, a selector valve, and an actuating cylinder. The hydraulic reservoir is a tank or a container designed to store sufficient hydraulic fluid for all conditions of operation. It replenishes the system fluid when needed and provides room for thermal expansion. Reservoirs in hydraulic systems which require a reserve of fluid for the emergency operation are equipped with standpipes in the bottom to retain the amount of fluid required for emergency operation. During normal operation, fluid is drawn through the standpipe. When system fluid is lost, emergency fluid is draw from the bottom of the reservoir.

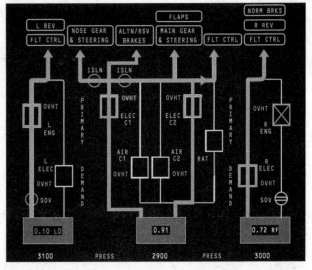

The engine driven pump which is directly coupled to the engine will operate when the engine is running. The pump draws fluid from the reservoir and expels it to the system through the filter and a check valve and the system pressure can be built up.

A hand pump is the simplest pump which is operated manually for emergency use in some light airplane. An electrically driven pump may be used as an emergency pump in larger aircraft. The engine driven pumps are used in main hydraulic power systems.

A filter is a screening or straining device used to clean the hydraulic fluid, thus preventing foreign particles and contaminating substances from remaining in the system. If such material is not removed, it may cause the entire hydraulic system of the aircraft to fail through the breakdown or malfunctioning of a single unit of the system.

Hydraulic pressure must be regulated in order to use it to perform the desired tasks. The pressure regulator unloads or relieves the engine driven pump when the desired pressure in the system is reached. Thus it is often referred to as an unloading valve. When one of the actuating units has been operated and pressure in the line between the pump and selector valve builds up to the desired value, a valve in the pressure regulator automatically opens and fluid is bypassed back to the reservoir. In the event of a malfunction of the pressure regulator, the relief valve (safety valve) will release any pressure above that maintained by the system pressure regulator.

The hydraulic system may be affected by technical problems or outside damage such as birdstrike. Problems with the hydraulic system can lead to problems with aircraft control and maneuverability. They may cause limited maneuverability (bank angle/turns), limited or no flap setting, limited bank angle (15degrees), higher speed and manual gear extension with no retraction. They may result in complete or partial failure of flight controls, gear extension, brakes, flaps, nose wheel steering. Flaps, ailerons, elevators, rudder, lift and roll spoilers may be out of use or be used with difficulty only.

Hydraulic failures may occur in different situations. In most cases there is no time pressure. However, depending on the aircraft type, individual problems may arise. Therefore, the pilot will normally inform ATC about controllability of the aircraft and his further intentions. Besides, the flight crew need time to check alternate systems and all other related functions and need more time than usual for actions such as manual gear extension. (643 words)

NEW WORDS

hydraulic	/haiˈdrɔːlik/	*adj.*	液压的
unit	/ˈjuːnit/	*n.*	组件
aviation	/ˌeiviˈeiʃən/	*n.*	航空
brake	/breik/	*n.*	刹车
sophisticated	/səˈfistiˌkeitid/	*adj.*	复杂的
wheel	/wiːl/	*n.*	机轮；盘
reservoir	/ˈrezəvwɑː/	*n.*	(液压)油箱
pump	/pʌmp/	*n.*	泵
filter	/ˈfiltə/	*n.*	过滤器
accumulator	/əˈkjuːmjuleitə/	*n.*	蓄压器
container	/kənˈteinə/	*n.*	容器
replenish	/riˈpleniʃ/	*v..*	更新；补充
thermal	/ˈθəːməl/	*adj.*	热的
reserve	/riˈzəːv/	*n.*	储备；备份
standpipe	/ˈstændˌpaip/	*n.*	竖管
expel	/ˌiksˈpel/	*v.*	挤出
couple	/ˈkʌpl/	*v.*	耦合
charge	/tʃɑːdʒ/	*v.*	充压，充电
piston	/ˈpistən/	*n.*	活塞
predetermine	/ˌpriːdiˈtəːmin/	*v.*	预定
outlet	/ˈaʊtlet/	*n.*	出口
spring	/spriŋ/	*n.*	弹簧
resistance	/riˈzistəns/	*n.*	阻力；阻滞
bypass	/ˈbaiˌpɑːs/	*v.*	旁通
screen	/skriːn/	*n. & v.*	过滤；屏幕
particle	/ˈpɑːtikl/	*n.*	颗粒
breakdown	/ˈbreikˌdaʊn/	*n.*	故障

malfunction	/mælˈfʌŋkʃən/	*n.*	故障
partial	/ˈpɑːʃəl/	*adj.*	部分的
alternate	/ɔːlˈtəːnit/	*adj.*	备用的
	/ˈɔːltənit/	*v.*	交替
		n.	备降场
strip	/strip/	*n.*	条；带
hand-over	/hændˈəuvə/	*n.*	移交
tow	/təu/	*v.*	牵引
aerodrome	/ˈeərəˌdrəum/	*n.*	机场
crosswind	/ˈkrɒswind/	*n.*	侧风；二边
component	/kəmˈpəunənt/	*n.*	部件；矢量
visibility	/ˌviziˈbiliti/	*n.*	能见度
ceiling	/ˈsiːliŋ/	*n.*	云高；升限
controllability	/ˌkəntrəuləˈbiliti/	*n.*	操纵性
intention	/ˌinˈtenʃən/	*n.*	意图

EXPRESSIONS

nose wheel steering　前轮转向系统　　　control surface　操纵面

relief valve　安全活门　　　　　　　　check valve　单向活门

actuating cylinder　动作筒　　　　　　thermal expansion　热膨胀

return line　回油管　　　　　　　　　hand pump　手动泵

NOTES

POB	persons on board	机上人员	NM	nautical mile	海里
NAV	navigation	导航	ACFT	aircraft	飞机；航空器
ILS	instrument landing system	仪表着陆系统			

EXERCISES

Comprehension of Text A

I. Describe the given aviation terms from Text A in English.

1. hydraulic power
2. engine driven pump
3. hydraulic reservoir
4. hydraulic pressure regulator

5. relief valve   6. controllability

II. Answer the following questions after you have read Text A.

1. Which systems are operated by the hydraulic system?

2. What are the basic components that comprise hydraulic system?

3. What does a standpipe in the hydraulic reservoir serve to do?

4. Does engine driven pump produce hydraulic pressure? Why?

5. What is the basic function of a check valve in the hydraulic system?

6. How is hydraulic pressure regulated in the system?

7. What are the possible reasons for hydraulic problems?

8. What are the possible consequences for hydraulic failure?

9. What should a pilot say when there is a hydraulic problem?

10. What services should ATC provide for an aircraft with hydraulic problem?

Reading Aloud

III. Read the following paragraph aloud until you can say it in a natural way from your memory.

Wheel Well Fire   (轮舱起火)

There are some reasons that may contribute to wheel fire. Overheated brakes can contribute to a wheel well fire if they become hot enough to ignite the tires, hydraulic fluid or other flammable(易燃的) substances. However, analysis by Boeing indicates that overheated brakes alone can not generate enough heat to trigger the wheel well fire detection system.

Should wheel well fire take place, some actions should be taken by pilots. In most cases extending the gear to cool the brakes and wheel well is effective at coping with the fire condition. Even so there have been a few incidents where extending the gear did not resolve the problem. One such example was an incident where the electrical lead to a hydraulic pump shorted and burned a small hole through the hydraulic pressure line and ignited the spraying fluid. After the gear was extended, the fire warning went out but the fluid from the hydraulic line continued to burn.

Wheel well fire indications should be treated with the same respect as other fire indications. This includes the requirement to land at the nearest suitable airport. The landing gear should be left extended unless it is absolutely necessary to retract it in order to reach for performance considerations. (205 words)

Vocabulary Practice

IV. Complete the following short passage by filling in the blanks with the words given in the box.

> warnings   recycling   landing   runways   installation   exterior
> models   flat   well   proximity

Upon arrival in Québec, when attempting to extend the landing gear, the flight crew received visible and audible 1_____ that the nose gear was not in the down and locked position. The checklist was completed and a second attempt at 2_____the landing gear was made without success. The landing gear manual release handle was activated twice without satisfactory results. The cabin was prepared for a nose gear retracted 3_____. The aircraft then landed on its nose.

The aircraft can be used on various types of surfaces, including unpaved/gravel 4_____. In May 1988, the *Service aérien gouvernemental du Québec* installed a kit supplied by Bombardier for operation from unpaved/gravel runways. The kit includes, among other items, two gravel deflectors for 5_____ on the nose landing gear and other protective devices for the main landing gear. The deflectors are used to protect the aircraft 6_____ surfaces and engine against damage that can be caused by solid particles that are projected during take-offs and landings. Only 8 of 255 CL-600 7_____ that were built are equipped with this kit.

When recovering the aircraft, the nose wheel oleo pneumatic shock absorber (减震器) was found to be 8_____. The nose gear was partly extended, the right deflector was stuck in the nose landing gear 9_____, and the wheels had pivoted and were out of alignment. The nose gear torque links had dragged on the runway, which had worn flats on them at the apex. The nose landing gear doors and nose landing gear well structure were also damaged. Also, the 10_____ switches, located on each side of the nose gear leg, had been struck and damaged by the lips of the tires.

Reading for More

V. Read the following passage and answer the given questions briefly.

The Braking System (刹车系统)

All hydraulic brake systems operate on the same basic principle. When the operator moves a brake pedal or other brake operating control, the movement is transmitted to a master cylinder or to a power brake control valve from which fluid pressure is delivered through connecting lines to a brake assembly connected to a wheel or shaft whose movement is to be braked. The fluid pressure acting on the brake assembly pushes brake linings into contact with surfaces of a rotating disc. The resulting friction slows – and finally stops – the continued rotation of the wheel or shaft to which the disc is connected. When the brake pedal or brake control is returned to the off position, brake operating pressure is relieved, the brake lining loses contact with the disc, and the wheel or shaft is free to turn again.

Aircraft wheel brake systems are dual in nature in that they are composed of two identical subsystems that can be operated independently of each other to provide separate braking action for the landing gear on each side of the aircraft. Each subsystem is operated by a brake pedal that is hinge-mounted to the top of the aircraft rudder pedal. Since each brake pedal can be operated independently, the brakes can be used for steering the aircraft. A list of components, which may be found in varying combinations to make up the different wheel brake systems, includes the following: master cylinder (or a power brake control valve), wheel brake assemblies de-boosters, parking brake valves, shuttle valves, accumulators, connecting lines, and bottles charged with compressed air. The minimum number of parts which could be used to perform the function of a simple wheel brake system are a master cylinder (or a power brake control valve), a wheel brake assembly, and connecting lines. (300 words)

Question 1　How does the braking system work?

Question 2　How many sub-systems does the aircraft wheel brake system have?

Question 3　What are the basic components of the aircraft wheel brake system?

VI. In this section, there are two passages. After you have read the two passages, there are five questions followed by four choices marked A, B, C and D. Decide which one is the most appropriate answer.

Passage One

The Landing Gear

The landing gear forms the principal support of the airplane on the surface. The most common type of landing gear consists of wheels, but airplanes can also be equipped with floats (浮筒) for water operations, or skis (雪橇) for landing on snow.

The landing gear on small airplanes consists of three wheels, two main wheels, one located on each side of the fuselage, and a third wheel, positioned either at the front or rear of the airplane. Landing gear employing a rear-mounted wheel is called a conventional landing gear (后三点式起落架). Airplanes with conventional landing gear are often referred to as tailwheel airplanes. When the third wheel is located on the nose, it is called a nosewheel, and

the design is referred to as a tricycle gear. A steerable nosewheel or tailwheel permits the airplane to be controlled throughout all operations while on the ground.

A tricycle gear (前三点式) airplane has three main advantages. It allows more forceful application of the brakes during landings at high speeds without resulting in the airplane nosing over (拿大顶). It permits better forward visibility for the pilot during takeoff, landing, and taxiing. Moreover, it tends to prevent ground looping (打地转) by providing more directional stability during ground operation since the airplane's center of gravity is forward of the main wheels. The forward CG, therefore, tends to keep the airplane moving forward in a straight line rather than ground looping.

Nosewheels are either steerable (可操纵的) or castering (转向的). Steerable nosewheels are linked to the rudders by cables or rods, while castering nosewheels are free to swivel (旋转). In both cases, you steer the airplane using the rudder pedals. However, airplanes with a castering nosewheel may require you to combine the use of the rudder pedals with independent use of the brakes.

On tailwheel airplanes, two main wheels, which are attached to the airframe ahead of its center of gravity, support most of the weight of the structure, while a tailwheel at the very back of the fuselage provides a third point of support. This arrangement allows adequate ground clearance for a larger propeller and is more desirable for operations on unimproved fields.

The main drawback (缺陷) with the tailwheel landing gear is that the center of gravity is behind the main gear. This makes directional control more difficult while on the ground. If you allow the airplane to swerve while rolling on the ground at a speed below that at which the rudder has sufficient control, the center of gravity will attempt to get ahead of the main gear. This may cause the airplane to ground loop. (425 words)

1. On the ground, the majority of the weight of an aircraft is supported by _____.

    A. the floats   B. the skis   C. the landing gear   D. the nose wheel

2. According to the passage, tailwheel and nosewheel are similar in that _____.

    A. they may be controllable   B. they are rear-mounted

    C. they are forward mounted   D. they are of tricycle type

3. A tricycle gear has more directional stability because _____.

    A. it can prevent the aircraft from nosing over

    B. the CG is forward of the main gears

    C. it may result in ground looping during turns

    D. the pilots have better visibility in the cockpit

4. Either steerable or castering, the nose wheel is controlled by _____.

    A. rudder pedals and the brakes   B. the brakes

C. rudder pedals or the brakes   D. rudder pedals
5. During ground movement, the tailwheel airplane has the primary risk of _____.
    A. directional control   B. ground clearance
    C. ground loop   D. rudder control

Passage Two

Faced with a Gear Problem

Carrying 140 passengers and six crew, the Airbus A320-232 aircraft departed from Burbank at 3:17 pm PDT. It was scheduled to fly 2,465 miles to JFK airport.

After takeoff from Burbank, the pilots realized that they could not retract the landing gear. They then flew low over Long Beach Municipal Airport in Long Beach to allow officials in the airport's control tower to take stock of the damage to its landing gear before attempting a landing, and it was found that the nosewheel was rotated ninety degrees to the left, perpendicular (垂直的) to the direction of the fuselage.

Rather than land at Long Beach Airport, the pilot-in-command took the decision that the aircraft would divert (备降) and land at Los Angeles International Airport, in order to take advantage of its long, wide runways and modern safety equipment.

The pilots flew the aircraft, which can carry up to 46,860 pounds of aviation fuel, for more than two hours in order to burn fuel and lower the risk of fire upon landing. This also served to lighten the plane, reducing potential stress on the landing gear and dramatically lowering landing speed as well. The Airbus A320 does not have the mechanical facility to dump fuel, despite various news agencies reporting that the aircraft was doing so over the ocean.

Emergency services and fire engines were standing by. Although foam trucks were available, they were not used. The US FAA no longer recommends pre-foaming runways, chiefly due to concerns that it would deplete firefighting foam supplies which might later be needed to respond to a fire; it is also difficult to determine exactly where a runway should be foamed, and pre-foaming might also reduce the effectiveness of the aircraft's brakes, potentially causing it to slide off the runway.

The pilot did an outstanding job. He kept the plane on its rear tires as long as he could before he brought the nose gear down. When the nose gear did touch down, there were sparks and flames from it, but no apparent damage to the rest of the plane. At 6:20pm, the aircraft came to a stop very close to the end of the 11,096-foot runway 25L. In an attempt to keep the nose gear off the ground as long as possible, reverse thrust was not used to slow the aircraft. The pilots therefore used a much larger portion of the available runway than in a typical

landing, stopping 1,000 feet/305 m before the end of the runway, validating the decision to divert from Long Beach, where the longest runway is 10,000 feet. (432 words)

6. The flight over the Long Beach Municipal Airport was attempted to_____.

   A. make a visual landing at the airport

   B. find out the condition of the nose gear

   C. try retracting the nose gear by impact

   D. rotate the nose gear to about 90 degrees

7. The flight was diverted to Los Angeles International Airport because _____.

   A. the pilot thought better of the airport conditions

   B. more fuel would be consumed with longer range

   C. the risk of wheel fire would be avoided on landing

   D. stress on landing gears on landing can be reduced

8. The news agencies seemed not to have the knowledge that _____.

   A. the Airbus A320 requested foaming the runway

   B. the Airbus A320 was heavier than expected

   C. less landing speed was more important then

   D. the Airbus A320 could not make fuel dumping

9. Which of the following statements is not the reason for not foaming the runway?

   A. Emergency services and fire engines.

   B. FAA pre-foaming recommendation.

   C. Difficulty determining proper foaming.

   D. Effectiveness of the aircraft's brakes.

10. The pilot of the Airbus 320 was greatly praised by the author because _____.

    A. the main gears touched down as late as possible

    B. the nose gear touched down as early as possible

    C. no damages were brought to the airplane on landing

    D. the nose gear was kept in the air as long as possible

VII. Read the following passage and retell it in your own words.

Hydraulic Fluid Leakage (液压油泄漏)

The first officer of AirTran Airways Flight 890 **identified**[1] and reported a leak from the right engine during a **post-flight**[2] inspection[3] at Atlanta. AirTran mechanics identified the source of the leak as a **chafed**[4] hydraulic pressure line to the right **thrust reverser**[5]. The AirTran maintenance controller in Orlando instructed

[1] 识别
[2] 飞行后
[3] 检查；巡查
[4] 受擦损的
[5] 反推装置

the mechanic to repair the leaking line and **deactivate**[6] the right [6] 断开
thrust reverser in accordance with AirTran's Minimum Equipment
List (MEL) procedures. However, instead of repairing the hydraulic
pressure line, the mechanics repaired the right engine hydraulic pump
case drain return line. The mechanics performed a leak check by
starting the **auxiliary power unit**[7] and turning on the electric [7] 辅助动力装置
hydraulic pumps to pressurize the airplane's hydraulic systems; no
leaks were detected.

Although the **mechanics**[8] were not required by company [8] 机务人员
procedures to test their repair by running the engines, this test would
have alerted the mechanics that they had incorrectly repaired the
hydraulic pump case drain line, which would have **overpressurized**[9] [9] 使……超压
the hydraulic pump and caused the hydraulic pump case seal to
rupture[10]. However, because the mechanics did not perform this test, [10] 破裂
the overpressure and rupture occurred during the airplane's climb out,
allowing depletion[11] of system A hydraulic fluid. [11] 耗尽

The crew **notified**[12] air traffic control that the airplane would be [12] 通知
returning to ATL(亚特兰大) and subsequently declared an emergency.
The flight crew's initial approach to the airport was high and fast
because of the workload associated with performing AirTran's
procedures for the loss of hydraulic system A and the limited amount
of time available to perform the procedures. Nevertheless, the crew
were able to **configure**[13] and stabilize the airplane for landing on [13] 建立形态
runway 09L. However, depletion of system A hydraulic fluid disabled
the nosewheel steering, **inboard**[14] flight spoilers, ground spoilers, [14] 内侧的
and left and right inboard brakes. The flight crew were able to land
the airplane using the left thrust reverser (the right thrust reverser was
fully **functional**[15] but intentionally deactivated by the mechanics), [15] 起作用的
outboard brakes (powered by hydraulic system B), and rudder. The
flight crew used the left thrust reverser and rudder in an attempt to
control the direction of the airplane down the runway, but use of the
rudder pedals in this manner had depleted the system A accumulator
pressure, which would have allowed three emergency brake
applications. The use of the right outboard brake without the right
inboard brake at a higher-than-normal speed and with heavy **gross
weight**[16] used up the remaining friction material on the right [16] 总重
outboard brake, causing it to fail. The lack of **brake friction**[17] [17] 摩擦
material on the right outboard brake caused one of the right outboard
brake pistons to over-travel, allowing system B hydraulic fluid to leak
out; as a result, the left outboard brake also failed. Loss of the left and
right inboard and outboard brakes, loss of nosewheel steering, and use
of asymmetric reverse thrust caused the flight crew to lose control of
the airplane. (485 words)

Text B

Passage Writing

VIII. Read the following passage carefully and write a passage under the given title with no less than 200 words.

A Gear Problem

_____

_____

Landing Gear Malfunction

Once the pilot has confirmed that the landing gear has in fact malfunctioned, and that one or more gear legs refuses to respond to the conventional or alternate methods of gear extension contained in the POH, there are several methods that may be useful in attempting to force the gear down. One method is to dive the airplane (in smooth air only) to VNE speed (red line on the airspeed indicator) and (within the limits of safety) execute a rapid pull up. In normal category airplanes, this procedure will create a 3.8 G load on the structure, in effect making the landing gear weigh 3.8 times normal. In some cases, this may force the landing gear into the down and locked position. This procedure requires a fine control touch and good feel for the airplane. The pilot must avoid exceeding the design stress limits of the airplane while attempting to lower the landing gear. The pilot must also avoid an accelerated stall and possible loss of control while attention is directed to solving the landing gear problem.

Another method that has proven useful in some cases is to induce rapid yawing. After stabilizing at or slightly less than maneuvering speed, the pilot should alternately and aggressively apply rudder in one direction and then the other in rapid sequence. The resulting yawing action may cause the landing gear to fall into place.

If all efforts to extend the landing gear have failed, and a gear up landing is inevitable, the pilot should select an airport with crash and rescue facilities. The pilot should not hesitate to request that emergency equipment be standing by.

When selecting a landing surface, the pilot should consider that a smooth, hard-surface runway usually causes less damage than rough, unimproved grass strips. A hard surface does, however, create sparks that can ignite fuel. If the airport is so equipped, the pilot can request that the runway surface be foamed. The pilot should consider burning off excess fuel. This will reduce landing speed and fire potential.

If the landing gear malfunction is limited to one main landing gear leg, the pilot should consume as much fuel from that side of the airplane as practicable, thereby reducing the weight of the wing on that side. The reduced weight makes it possible to delay the unsupported wing from contacting the surface during the landing roll until the last possible moment. Reduced impact speeds result in less damage.

If only one landing gear leg fails to extend, the pilot has the option of landing on the available gear legs, or landing with all the gear legs retracted. Landing on only one main gear usually causes the airplane to veer strongly in the direction of the faulty gear leg after touchdown. If the landing runway is narrow, and/or ditches and obstacles line the runway edge, maximum directional control after touchdown is a necessity. In this situation, a landing with all three gear retracted may be the safest course of action.

If the pilot elects to land with one main gear retracted, the landing should be made in a nose-high attitude with the wings level. As airspeed decays, the pilot should apply whatever aileron control is necessary to keep the unsupported wing airborne as long as possible. Once the wing contacts the surface, the pilot can anticipate a strong yaw in that direction. The pilot must be prepared to use full opposite rudder and aggressive braking to maintain some degree of directional control.

In case of landing with a retracted nosewheel, the pilot should hold the nose off the ground until almost full up-elevator has been applied. The pilot should then release back pressure in such a manner that the nose settles slowly to the surface. Applying and holding full up-elevator will result in the nose abruptly dropping to the surface as airspeed decays, possibly resulting in aircraft damage. Brake pressure should not be applied during the landing roll unless absolutely necessary to avoid a collision with obstacles. If the landing must be made with only the nose gear extended, the initial contact should be made on the aft fuselage structure with a nose-high attitude. This procedure will help prevent wheel problems. The pilot should then allow the nosewheel to gradually touch down, using nosewheel steering as necessary for directional control. (718 words)

NEW WORDS

confirm	/kən'fəːm/	v.	证实；确认
dive	/daiv/	v.	俯冲

execute	/'eksikjuːt/	v.	执行
accelerate	/æk'seləreit/	v.	加速
induce	/in'djuːs/	v.	诱发
aggressively	/əg'resivli/	adj.	攻击性地；过量地
sequence	/'siːkwəns/	n.	顺序
inevitable	/in'evitəbl/	a.	必然的
crash	/kræʃ/	v.	坠毁
rescue	/'reskjuː/	n. &. v.	营救
rough	/rʌf/	a.	粗糙的
spark	/spɑːk/	n.	火花
ignite	/ig'nait/	v.	点燃
contact	/'kɒntækt/	n.	接触；联系
impact	/'impækt/	n.	撞击；影响
excess	/ik'ses/	n.	过多
foam	/fəum/	n. &. v.	泡沫；铺设泡沫
potential	/pə'tenʃəl/	n.	潜力；可能
		adj.	潜在的；可能的
leg	/leg/	n.	段；边
extend	/iks'tend/	v.	伸出；放出
veer	/viə/	v.	改变
faulty	/'fɔːlti/	adj.	故障的
ditch	/ditʃ/	n. &. v.	壕沟；水上迫降
retract	/ri'trækt/	v.	收上
decay	/di'kei/	v.	消减
anticipate	/æn'tisipeit/	v.	预期；预想
release	/ri'liːs/	v.	松开；释放
abruptly	/əb'rʌptli/	adv.	突然地

EXPRESSIONS

POM   Pilot Operation Manual   飞行操作手册

Lesson 6

Automatic Flight & Instruments

Preview

Automatic flight control systems are one of the most crucial systems in modern aircraft which are related to many other aircraft systems such as flight management systems, air data systems, flight control systems. Text A helps you acquire some fundamental knowledge as to automatic flight. Text B specifies the Flight Management System.

Text A

Warming-up Activities

Picture Description

Please describe the following picture and be prepared to answer some questions.

Relevant Questions

1. Where can you find the display unit in the cockpit?
2. What kinds of information are shown in different parts of the display unit?
3. How is this display related to the automatic flight system?
4. What do you think are the advantages of PFD in comparison with the traditional Attitude Direction Indicator (ADI)?

Automatic Flight

The automatic flight control systems (AFCS) consist of three independent systems: digital flight control system (DFCS), yaw damper system, and autothrottle system. These systems provide automatic airplane stabilization about the pitch, roll, and yaw axes and control the airplane with selective guidance from radio, heading, flight management computer, and air data computer inputs.

The DFCS system is a two-axis (pitch and roll) system which operates the elevators and ailerons to automatically maintain altitude, airspeed and/or guide the airplane to designated locations and make automatic landings. Control functions are also translated into flight director commands for display on the pilots' attitude director indicators, thereby providing the pilots' flight attitude commands during manual operation or allowing the pilots to monitor autopilot operation. Automatic stabilizer trimming relieves sustained elevator loads which might be incurred due to fuel burnoff.

The yaw damper systems operate the rudders to correct any periodic yaw oscillations (dutch roll). The yaw damper system is a full-time, series-connected, stability augmentation system. Airplane periodic oscillations are detected with a rate sensor in a yaw damper computer. The rudder is displaced at the proper time to dampen out any dutch roll before it can significantly affect the flight path of the airplane. The yaw damper system and actuators are connected such that no rudder feedback is applied to the pedals, thereby allowing the system to operate independently without interfering with pilot initiated rudder commands.

An autothrottle system automatically maintains selected airspeeds or Mach during cruise conditions and maintains selected engine thrust settings when making flight director controlled take-offs or autopilot/flight director controlled landing approaches by adjusting engine thrust levers.

The autothrottle system automatically positions all thrust levers to maintain a computed engine thrust level during take-off or go-around, and a selected or FMC computed airspeed (IAS or Mach) during cruise. This system also computes and maintains a safe airspeed during holding patterns and initial approach maneuvers. The autothrottle system also retards the thrust levers during automatic landing during flare. The autothrottle system consists of a computer which operates the thrust levers through integrated servo assemblies.

The Digital Flight Control System (DFCS) incorporates two separate channels (A, B). Each channel controls the pitch and roll axes and provides mach trim and speed trim control. Flight director commands and flag logic are connected to the captain's ADI (system A) and the first officer's ADI (system B). DFCS control of the airplane is enabled by engaging channel A or B switch on DFCS mode select panel. When dual channel option is provided, channels A and B can be engaged simultaneously when making fail-passive automatic landings.

The Mach Trim System provides automatic repositioning of the elevators as a function of Mach number. As the airplane enters the Mach tuck region, the elevator is repositioned to provide a new neutral in an upward direction which is proportional to the increase in Mach. The Mach Trim System operates with or without the autopilot system engaged. The Mach Trim System operates in conjunction with the elevator hydraulic power control units and stabilizer/elevator neutral shift mechanism. Each flight control computer (FCC) includes a Mach trim computer with automatic switching to other system if failure occurs.

The speed trim system actuates the stabilizer trim to maintain positive speed control. The system may be required during takeoff or go-around with low gross weight, aft CG and high thrust. The system is operational when autopilot is not engaged, stabilizer is not trimming, and control column displacement switches are not actuated. Each FCC contains speed trim computer with automatic monitoring and automatic switching between systems (when no faults). The monitoring automatically selects the operational channel when a single channel failure occurs.

Each DFCS channel consists of an FCC, an ADI, and two A/P actuators. A mode selector panel controls autopilot/flight director systems and the autothrottle system. An AFC accessory unit and automatic stabilizer trim servo are also installed as well as elevator spoiler, flap position, and stabilizer position sensors.

Autopilot/flight director disengage warning consists of autopilot warning lights and wailer operation when certain autopilot functions are not correct and/or the autopilot is disengaged. ADI pitch and/or roll bars are biased out of view and/or the flight director flag (or computer flag) is biased into view as an indication of certain flight director malfunctions. (709 words)

NEW WORDS

digital	/'dɪdʒɪtəl/	adj.	数字的
damper	/'dæmpə/	n.	阻尼器
autothrottle	/ˌɔːtə'θrɒtl/	n.	自动油门
axis	/'æksɪs/	n.	轴
input	/'ɪnput/	n.& v.	输入
electronic	/ɪlek'trɒnɪk/	adj.	电子的
relieve	/rɪ'liːv/	v.	减轻
sustain	/sə'steɪn/	v.	持续
incur	/ɪn'kə:/	v.	导致
augmentation	/ˌɔːgmen'teɪʃən/	n.	增加；增益
detect	/dɪ'tekt/	v.	探测
displace	/dɪs'pleɪs/	v.	位移
path	/pɑːθ/	n.	轨迹
feedback	/'fiːdbæk/	n.	反馈
initiate	/ɪ'nɪʃɪeɪt/	v.	开始；启用
thrust	/'θrʌst/	n.	推力
lever	/'levə/	n.	手柄
maneuver	/mə'nuːvə/	n.	机动
assembly	/ə'semblɪ/	n.	装置
servo	/'səːvə/	n.	伺服机构
channel	/'tʃænəl/	n.	信道，通道
logic	/'lɒdʒɪk/	n.	逻辑

dual	/'djuəl/	*adj.*	双重的；双的
tuck	/tʌk/	*v.*	收上
neutral	/'njuːtrəl/	*adj.*	中立（位）的
engage	/in'geidʒ/	*v.*	接通
positive	/'pɒzitiv/	*adj.*	正的，积极的
bar	/bɑː/	*n.*	指令杆

NOTES

AFCS	Automatic Flight Control System	自动飞行控制系统
ADC	Air Data Computer	大气数据计算机
DFCS	Digital Flight Control System	数字飞行控制系统
ADI	attitude director indicator	姿态指引仪
A/T	autothrottle	自动油门
A/P	autopilot	自动驾驶仪
FCC	flight control computer	飞行控制计算机
IAS	Indicated Airspeed	指示空速
CG	Center of Gravity	重心
FD	Flight Director	飞行指引仪

EXERCISES

Comprehension of Text A

I. Describe the given aviation terms from Text A in English.

1. DFCS

2. autothrottle

3. yaw damper

4. Mach tuck region

5. autopilot

6. flight director

II. Answer the following questions after you have read Text A.

1. How many subsystems do automatic flight control systems have?

2. What kinds of navigation information are needed for AFCS ?

3. What is the basic function of the DFCS system?

4. How is the DFCS displayed on the EADI?

5. Do you know anything about the yaw damper system?

6. In what conditions can the autothrottle system be used?

7. How do the DFCS channels work?

8. Can you describe the operation principle of the Mach Trim system?

9. What components does each DFCS channel have?

10. When does the Autopilot/flight director disengage warning occur?

Reading Aloud

III. Read the following paragraph aloud until you can say it in a natural way from your memory.

Flight Instruments

Flight instruments in the cockpit of an aircraft provide pilots with information about the flight situation of that aircraft, such as height, speed and attitude. The flight instruments are of particular use in conditions of poor visibility, such as in cloud, when such information is not available from visual reference outside the aircraft. The main flight instruments are altimeter, attitude indicator, and airspeed indicator.

The altimeter shows the aircraft's height above some reference level by measuring the local air pressure. It is adjustable for local barometric pressure which must be set correctly to obtain accurate altitude readings.

The attitude indicator shows the aircraft's attitude relative to the horizon. From this the pilot can tell whether the wings are level and if the aircraft nose is pointing above or below the horizon. This is a primary instrument for instrument flight and is also useful in conditions of poor visibility. Pilots are trained to use other instruments in combination should this instrument or its power fail.

The airspeed indicator shows the aircraft's speed relative to the surrounding air. It works by measuring the ram-air pressure in the aircraft's pitot tube. The indicated airspeed must be corrected for air density in order to obtain the true airspeed, and for wind conditions in order to obtain the speed over the ground. (217 words)

Vocabulary Practice

IV. Complete the following short passage by filling in the blanks with the words given in the box.

relative	banked	scale	mounted	horizon
gyro	displays	cruising	align	exceeded

The Attitude Director Indicator

The attitude indicator, with its miniature airplane and horizon bar, 1_____a picture of the attitude of the airplane. The relationship of the miniature airplane to the horizon bar is the same as the relationship of the real airplane to the actual 2_____. The instrument gives an instantaneous indication of even the smallest changes in attitude.

The gyro in the attitude indicator is 3_____on a horizontal plane and depends upon rigidity in space for its operation. The horizon bar represents the true horizon. This bar is fixed to the 4_____and remains in a horizontal plane as the airplane is pitched or banked about its lateral or longitudinal axis, indicating the attitude of the airplane 5_____to the true horizon.

An adjustment knob is provided with which the pilot may move the miniature airplane up or down to 6_____ the miniature airplane with the horizon bar to suit the pilot's line of vision. Normally, the miniature airplane is adjusted so that the wings overlap the horizon bar when the airplane is in straight-and-level 7_____flight.

The pitch and bank limits depend upon the make and model of the instrument. Limits in the banking plane are usually from 100° to 110°, and the pitch limits are usually from 60° to 70°. If either limit is 8_____, the instrument will tumble or spill and will give incorrect indications until restabilized. A number of modern attitude indicators will not tumble. Every pilot should be able to interpret the banking scale. Most banking scale indicators on the top of the instrument move in the same direction from that in which the airplane is actually 9_____.

Some other models move in the opposite direction from that in which the airplane is actually banked. This may confuse the pilot if the indicator is used to determine the direction of bank. This 10_____should be used only to control the degree of desired bank. The relationship of the miniature airplane to the horizon bar should be used for an indication of the direction of bank. The attitude indicator is reliable and the most realistic flight instrument on the instrument panel. Its indications are very close approximations of the actual attitude of the airplane. (371 words)

Reading for More

V. Read the following passage and answer the given questions briefly.

Blocked Pitot System

The pitot system can become blocked completely or only partially if the pitot tube drain hole remains open. If the pitot tube becomes blocked and its associated drain hole remains clear, ram air no longer is able to enter the pitot system. Air already in the system will vent through the drain hole, and the remaining pressure will drop to ambient (outside) air pressure.

Under these circumstances, the airspeed indicator reading decreases to zero, because the airspeed indicator senses no difference between ram and static air pressure. The airspeed indicator acts as if the airplane were stationary on the ramp. The apparent loss of airspeed is not usually instantaneous. Instead, the airspeed will drop toward zero.

If the pitot tube, drain hole, and static system all become blocked in flight, changes in airspeed will not be indicated, due to the trapped pressures. However, if the static system remains clear, the airspeed indicator acts as an altimeter. An apparent increase in the ram air pressure relative to static pressure occurs as altitude increases above the level where the pitot tube and drain hole became blocked. This pressure differential causes the airspeed indicator to show an increase in speed. A decrease in indicated airspeed occurs as the airplane descends below the altitude at which the pitot system becomes blocked.

The pitot tube may become blocked during flight through visible moisture. Some airplanes may be equipped with pitot heat for flight in visible moisture. Consult the AFM or POH for specific procedures regarding the use of pitot heat. (255 words)

Question 1   What may happen if the pitot system becomes blocked?

Question 2   What are the reasons for the blockage of pitot tube?

Question 3   If the pitot tubes become blocked in flight, while the static system remains clear, what may be caused by the pressure differential?

VI. In this section, there are two passages. After you have read the two passages, there are five questions followed by four choices marked A, B, C and D. Decide which one is the most appropriate answer.

Passage One

Pitot-Static System

The source of the pressure for operating the airspeed indicator, the vertical speed indicator, and the altimeter is the pitot-static system. The major components of the pitot-static system are the impact pressure chamber (膜盒) and lines, and the static pressure chamber and lines, each of which are subject to total or partial blockage by ice, dirt, and/or other foreign matter. Blockage of the pitot-static system will adversely affect instrument operation.

Partial static system blockage is dangerous in that it may go unrecognized until a critical phase of flight. During takeoff, climb, and level-off at cruise altitude the altimeter, airspeed indicator, and vertical speed indicator may operate normally. No indication of malfunction may be present until the airplane begins a descent.

If the static reference system is severely restricted, but not entirely blocked, as the airplane descends, the static reference pressure at the instruments begins to lag (滞后) behind the actual outside air pressure. While descending, the altimeter may indicate that the airplane is higher than actual because the obstruction slows the airflow from the static port to the altimeter. The vertical speed indicator confirms the altimeter's information regarding rate of change, because the reference pressure is not changing at the same rate as the outside air pressure. The airspeed indicator, unable to tell whether it is experiencing more airspeed pitot pressure or less static reference pressure, indicates a higher airspeed than actual. To the pilot, the instruments indicate that the airplane is too high, too fast, and descending at a rate much less than desired.

If the pilot levels off and then begins a climb, the altitude indication may still lag. The vertical speed indicator will indicate that the airplane is not climbing as fast as actual. The indicated airspeed, however, may begin to decrease at an alarming rate. The least amount of pitch-up attitude may cause the airspeed needle to indicate dangerously near stall speed.

Managing a static system malfunction requires that the pilot know and understand the airplane's pitot-static system. If a system malfunction is suspected, the pilot should confirm it by opening the alternate static source. This should be done while the airplane is climbing or descending. If the instrument needles move significantly when this is done, a static pressure problem exists and the alternate source should be used during the remainder of the flight. (386 words)

1. The main components of the pitot-static system can be affected by _____.
   A. the source of the air pressure
   B. the impact pressure chamber
   C. abnormal instrument operation
   D. blockage caused by foreign matters
2. Why is partial static system blockage very dangerous?
   A. Because it may have some impacts on the impact pressure.
   B. Because it will adversely affect instrument operation.
   C. Because the pilot may not realize it until crucial period.
   D. Because all pitot-static instruments may operate normally.
3. Instrument indications may lag behind existing outside air when there is _____.
   A. partial blockage of static system during descent
   B. malfunction in the cockpit pitot-static instruments
   C. significant change in the outside air pressure
   D. sudden rapid decrease in the aircraft's flight level
4. Which of following is *not* the reason for near stall speed indication?
   A. Change from level flight to climb.

B. Change in airplane pitch attitude.

C. Incorrect VSI indication.

D. Changes in airspeed warnings.

5. When the pilot is not sure of the normal operation of the pitot-static system, he should turn to _____.

    A. the indicated airspeed   B. system malfunction indication

    C. the standby static source   D. airplane's phase of flight

Passage Two

The Gyro Instruments

The gyro instruments include the heading indicator, attitude indicator, and turn coordinator (or turn-and-slip indicator). Each contains a gyro rotor driven by air or electricity and each makes use of the gyroscopic principles to display the attitude of the aircraft. It is important that instrument pilots understand the gyro instruments and the principles governing their operation.

The primary trait of a rotating gyro rotor is rigidity (定轴性) in space, or gyroscopic inertia. Newton's First Law states in part, "A body in motion tends to move in a constant speed and direction unless disturbed by some external force." The spinning rotor inside a gyro instrument maintains a constant attitude in space as long as no outside forces change its motion. This stability increases if the rotor has great mass and speed. Thus, the gyros in aircraft instruments are constructed of heavy materials and designed to spin rapidly (approximately 15,000 rpm for the attitude indicator and 10,000 rpm for the heading indicator).

The heading indicator and attitude indicator use gyros as an unchanging reference in space. Once the gyros are spinning, they stay in constant positions with respect to the horizon or direction. The aircraft heading and attitude can then be compared to these stable references. For example, the rotor of the universally mounted gyro remains in the same position even if the surrounding gimbals, or circular frames, are moved. If the rotor axis represents the natural horizon or a direction such as magnetic north, it provides a stable reference for instrument flying.

Another characteristic of gyros is precession (进动), which is the tilting or turning of the gyro axis as a result of applied forces. When a deflective force is applied to the rim of a stationary gyro rotor (转子), the rotor moves in the direction of the force. When the rotor is spinning, however, the same forces causes the rotor to move in a different direction, as though the force had been applied to a point 90° around the rim in the direction of rotation. This turning movement, or precession, places the rotor in a new plane of rotation, parallel to the applied force.

Air or electricity supplies the power to operate gyro instruments in light aircraft. If the directional indicator and attitude indicator are air-driven (as they generally are), the turn-and-slip indicator is electrically powered. The advantage of this arrangement is that if the vacuum system (which supplies air) fails, the instrument pilot still has the compass and the turn indicator for attitude and direction reference, in addition to the pitot-static instruments. (427 words)

6. Turn coordinator is cited in the first paragraph to show the importance of _____.
    A. the gyroscopic principles   B. the gyro's power supply
    C. the attitude of the aircraft   D. gyro instrument operation
7. Heavy materials are used to make gyros in flight instruments in order to _____.
    A. increase the weight of flight instruments
    B. increase rotation speed of the gyros
    C. change the status of spinning gyros
    D. heighten the stability of spinning gyros
8. Gyros in heading indicators work in relation to the horizon to provide_____.
    A. rigidity in space   B. faster spinning speed
    C. stable reference   D. visual reference
9. Difference in the direction of rotor motion occurs when _____.
    A. there is more stability in space
    B. a force acts on the spinning gyro
    C. axis of the gyro moves accordingly
    D. the new plane is manufactured
10. In light aircraft, air and electricity serve as powers for gyro instruments because _____.
    A. they two can backup each other
    B. should one fail, the other operates
    C. electricity has advantages over air
    D. the pilot may have more instruments

VII. Read the following passage and retell it in your own words.

Autopilot

Not all of the passenger aircraft flying today have an autopilot system. Older and smaller general aviation aircraft especially are still **hand-flown**[1], while small airliners with less than twenty seats may also be without an autopilot as they are used on **short-duration**[2] flights with two pilots. The fitment of autopilots to airliners with more than twenty seats is generally made **mandatory**[3] by international aviation regulations. There are three levels of control in autopilots for smaller aircraft. A single-**axis**[4] autopilot controls an aircraft in the roll axis only; such autopilots are also known **colloquially**[5] as "wing levelers", reflecting their limitations. A two-axis autopilot controls an aircraft in the pitch axis as well as roll, and may be little more than a "wing leveler" with limited pitch-oscillation-correcting ability; or it may receive inputs from on-board radio navigation systems to provide true automatic flight guidance once the aircraft has taken off until shortly before landing; or its capabilities may lie somewhere between these two extremes. A three-axis autopilot adds control in the yaw axis and is not required in many small aircraft.

Autopilots in modern complex aircraft are three-axis and generally divide a flight into taxi, takeoff, ascent, level, descent, approach and landing **phases**[6]. Autopilots exist that automate all of these flight phases except the taxiing. An autopilot-controlled landing on a runway and controlling the aircraft on rollout (i.e. keeping it on the centre of the runway) is known as a CAT IIIb landing or **Autoland**[7], available on many major airports' runways today, especially at airports subject to **adverse**[8] weather phenomena such as fog. Landing, rollout and taxi control to the aircraft parking position is known as CAT IIIc. This is not used to date but may be used in the future. An autopilot is often an integral component of a Flight Management System.

Modern autopilots use computer software to control the aircraft. The software reads the aircraft's current position, and controls a Flight Control System to guide the aircraft. In such a system, besides classic flight controls, many autopilots incorporate thrust control capabilities that can control throttles to **optimize**[9] the air-speed, and move fuel to different tanks to balance the aircraft in an optimal attitude in the air. Although autopilots handle new or dangerous situations **inflexibly**[10], they generally fly an aircraft with a lower **fuel consumption**[11] than a human pilot.

The autopilot in a modern large aircraft typically reads its position and the aircraft's attitude from an inertial guidance system. **Inertial**[12]

[1] 手动飞行
[2] 短程飞行
[3] 强制的
[4] 轴
[5] 口语地
[6] 阶段
[7] 自动着陆
[8] 不利的
[9] 使……最佳
[10] 不灵活的
[11] 油耗
[12] 惯性的

guidance systems accumulate errors over time. They will incorporate error
reduction systems so that any errors are **dissipated**[13] in different [13] 分散
directions and have an overall **nulling**[14] effect. Error in gyroscopes is [14] 归零
known as drift. This is due to physical properties within the system, be it
mechanical or **laser**[15] guided, that corrupt positional data. The [15] 激光
disagreements between the two are resolved with digital signal processing,
most often a six-dimensional Kalman filter. The six dimensions are usually
roll, pitch, yaw, altitude, latitude and longitude. Aircraft may fly routes that
have a required performance factor, therefore the amount of error or actual
performance factor must be monitored in order to fly those particular
routes. The longer the flight, the more error accumulates within the
system. Radio aids such as **DME**[16], DME updates and GPS may be used [16] 测距仪
to correct the aircraft position.(547 words)

Text B

Passage Rewriting

VIII. Read the following passage carefully and rewrite the passage under the given title with no less than 200 words.

The Adoption of FMS

The Flight Management System

A flight management system is a fundamental part of a modern aircraft in that it controls the navigation. The flight management system (FMS) is the avionics that holds the flight plan, and allows the pilot to modify as required in flight. The FMS uses various sensors to determine the aircraft's position. Given the position and the flight plan, the FMS guides the aircraft along the flight plan. The FMS is normally controlled through a small screen and a keyboard. The FMS sends the flight plan for display on the EFIS, Navigation Display (ND) or MultiFuction Display (MFD). The modern FMS was introduced on the Boeing 767, though earlier navigation computers did exist. Now, FMS exists on aircraft as small as the Cessna 172. In its evolution an FMS has many different sizes, capabilities and controls. However, certain characteristics are common to all FMS.

All FMS contain a navigation database. The navigation database contains the elements from which the flight plan is constructed. These are defined via the ARINC 424 standard. The navigation database is normally updated every 28 days, in order to ensure that its contents are current. Each FMS contains only a subset of the ARINC data, relevant to the capabilities of the FMS.

The navigation database contains all of the information required for building a flight plan and information relevant to it. These include waypoints, airways (highways in the sky), radio navigation aids including DME (Distance Measuring Equipment), VOR (VHF

Omni-directional Range) and NDB (Nondirectional beacons), airports, runways , Standard Instrument Departure (SID) , Standard Terminal Arrival (STAR), holding patterns and a variety of related and often installation specific information.

The flight plan is generally determined on the ground, before departure either by the pilot for smaller aircraft or a professional dispatcher for airliners. It is entered into the FMS either by typing it in, selecting it from a saved library of common routes (Company Routes) or via a datalink with the airline dispatch center.

During preflight, other information relevant to managing the flight plan is entered. This can include performance information such as gross weight, fuel weight, and center of gravity. It will include altitudes including the initial cruise altitude. For aircraft that do not have a GPS, the initial position is also required.

The pilot uses the FMS to modify the flight plan, in flight for a variety of reasons. Significant engineering design minimizes the keystrokes in order to minimize pilot workload in flight and eliminate any confusing information (Hazardously Misleading Information). The FMS also sends the flight plan information for display on the Navigation Display (ND) of the flight deck instruments (EFIS). The flight plan generally appears as a magenta line, with other airports, radio aids and waypoints displayed.

Special flight plans, often for tactical requirements including search patterns, in-flight refueling tanker orbits, calculated air release points (CARP) for accurate parachute jumps are just a few of the special flight plans some FMS can calculate.

Once in flight, a principal task of the FMS is to determine the aircraft's position and the accuracy of that position. Simple FMS uses a single sensor, generally GPS in order to determine position. But modern FMS uses as many sensors as they can in order to determine and validate exactly their position. Some FMS uses a Kalman filter to integrate the positions from the various sensors into a single position.

Inertial Reference Systems (IRS) use ring laser gyros and accelerometers in order to calculate the aircraft position. They are highly accurate and independent of outside sources. Airliners use the weighted average of three independent IRS to determine the "triple mixed IRS" position.

The FMS constantly crosschecks the various sensors and determines a single aircraft position and accuracy. Given the flight plan and the aircraft's position, the FMS calculates the course to follow. The pilot can follow this course manually (much like following a VOR radial), or the autopilot can be set to follow the course.

The FMS mode is normally called LNAV or lateral navigation for the lateral flight plan and VNAV or vertical navigation for the vertical flight plan. LNAV provides roll steering command to the autopilot and VNAV provides speed/altitude targets to the autopilot. (696

words)

NEW WORDS

avionics	/ˌeiviˈɒniks/	n.	航空电子
keyboard	/ˈkiːˌbɔːd/	n.	键盘
database	/ˈdeitəˌbeis/	n.	数据库
subset	/sʌbˈset/	n.	子集；部分
waypoint	/ˈweiˌpɔint/	n.	航路点
airway	/ˈeəwei/	n.	航路
dispatcher	/disˈpætʃə/	n.	签派员
route	/ruːt/	n.	航线
datalink	/ˈdeitəlink/	n.	数据联路
keystroke	/ˈkiːstrəuk/	n.	敲击键盘
magenta	/məˈdʒentə/	n.	洋红色
refuel	/ˌriːˈfjuəl/	v.	加油
parachute	/ˈpærəʃuːt/	n.	降落伞
accuracy	/ˈækjurəsi/	n.	精确度
validate	/ˈvælideit/	v.	使生效
laser	/ˈleizə/	n.	激光
accelerometer	/ækˌseləˈrɒmitə/	n.	加速计
gyro	/ˈdʒaiərəu/	n.	陀螺
crosscheck	/ˈkrɒstʃek/	n.	交叉检查
diameter	/daiˈæmitə/	n.	直径

NOTES

FMS	flight management system	飞行管理系统
MFD	MultiFuction Display	多功能显示器
CARP	calculated air release points	计算空气释放点
GPS	Global Positioning System	全球定位系统
IRS	Inertial Reference Systems	惯性基准系统
ANP	Actual Navigation Performance	实际导航性能
RNP	Required Navigation Performance	所需导航性能
LNAV	Lateral navigation	水平导航
VNAV	Vertical navigation	垂直导航

Lesson 7

Powerplant, APU & Electrical System

Preview

> Turbine engines, powering the majority of modern civil transport aircrafts, have the following advantages over reciprocating ones: less vibration, increased aircraft performance, reliability, and ease of operation. There are four types of turbine engines – turbojet, turboprop, turbofan, and turbo-shaft. All of them consist of an air inlet, compressor, combustion chambers, turbine section, and exhaust.

Text A

Warming-up Activities

Picture Description

Please describe the following picture and be prepared to answer some questions.

Relevant Questions

1. How many engines does a B747 have?
2. What are the possible damages caused by foreign object ingestion?
3. As a pilot, what will you do if you encounter engine fire in flight?
4. In which phase of flight is engine failure a greater threat to aviation safety? Why?

Powerplant

The B737 is powered by two CFM56-3 engines. The engine is a dual rotor axial flow turbofan. The N1 rotor consists of a fan, a three stage booster section connected by a shaft to a four stage low pressure turbine. The N2 rotor consists of a high pressure compressor and a high pressure turbine. The N1 and N2 rotors are mechanically independent. The N2 rotor drives the engine gearboxes. A bleed air powered starter motor is connected to the N2 rotor.

The main engine control (MEC) schedules fuel to provide the thrust called for by the forward thrust lever setting. The fuel flow is further refined electronically by the power management control (PMC) without moving the thrust levers. Each engine has individual flight deck controls. Thrust is set by positioning the thrust levers. The thrust levers are positioned automatically by the auto-throttle system or manually by the flight crew. The forward thrust levers control forward thrust from idle to maximum. With the PMCs ON or OFF, advancing the thrust levers full forward provides some over-boost and should be considered only during emergency situations when all other available actions have been taken and terrain contact is imminent. The reverse thrust levers control thrust from reverse idle to maximum reverse.

Certain engine malfunctions can result in airframe vibrations from the windmilling engine. As the airplane transitions from cruise to landing, there can be multiple, narrow regions of altitudes and airspeeds where the vibration level can become severe. In general, airframe vibrations can best be reduced by descending and reducing airspeed. However, if after descending and reducing airspeed, the existing vibration level is unacceptable, and if it is impractical to further reduce airspeed, the vibration level may be reduced to a previous, lower level by a slight increase in airspeed.

Engine indications are displayed on the center instrument panel by the Engine Instrument System (EIS). N1, EGT, N2, and FF/FU are the primary indications and are displayed as both digital readouts and round dial/moving pointer indications. N1, EGT, and N2 have operating and caution ranges and limits indicated by green and yellow bands and red radials. When the round red warning light above the indicator is illuminated, it indicates the limit for the engine parameter displayed below it has been reached or exceeded. The red warning light remains illuminated until the engine parameter is reduced below the limit.

Oil pressure and oil temperature indications are displayed with a round dial/moving pointer. Operating and caution ranges and limits are displayed with green and yellow bands and red radials. The oil quantity indicator displays a digital readout of quantity as a percentage of full. The airborne vibration monitor indications are displayed with a round dial/moving pointer. N1, N2, oil quantity, and engine vibration are displayed directly from the engine sensors.

The thrust control system consists of a hydro-mechanical MEC unit and a PMC unit mounted on each engine. The PMC is an electronic system with limited authority over the MEC. The PMC uses MEC power lever angle, N1 speed, inlet temperature and pressure to adjust, or trim, the MEC to obtain the desired N1 speed. The PMC adjusts fuel flow as a function of thrust lever angle. The PMC provides a constant thrust climb feature once the thrust lever is set at the beginning of climb. Thus, when thrust is set for the climb, the PMC automatically maintains that thrust throughout the climb profile with no further thrust lever adjustments. If the thrust lever is repositioned, the PMC maintains the setting corresponding to the new thrust lever angle. The PMC includes failure detection and annunciation modules which detect PMC failures and provide a signal to the crew. The PMC INOP light, the ENG system annunciator, and the MASTER CAUTION lights illuminate. For a PMC failure, the PMC can be selected OFF by a switch on the aft overhead panel. The engine speed is then controlled by the hydro-mechanical MEC only. The PMC INOP Light is suppressed below starter cutout engine speed.

Fuel is delivered under pressure from fuel pumps located in the fuel tanks. The fuel enters the engine through the fuel shutoff valve. The fuel passes through the first stage engine fuel pump where pressure is increased. It then passes through two fuel/oil heat exchangers where engine oil heats the fuel. A fuel filter then removes contaminants. Fuel automatically bypasses the filter if the filter becomes saturated. Before the fuel bypass occurs, the fuel FILTER BYPASS alert illuminates on the fuel control panel. The second stage engine fuel pump provides high pressure fuel to the main engine control (MEC). As the fuel leaves the second stage pump, a portion of the fuel is diverted to run the hydro-mechanical portion of the MEC. This fuel is filtered again and then routed through the fuel heater a second time. The MEC meters the correct amount of fuel to the combustor.

The engine fuel shutoff valve and MEC fuel shutoff valve allow fuel flow to the engine when both valves are open. The valves are open when the engine fire warning switch is in OFF and the start lever is in IDLE. Fuel flow is measured after the MEC fuel shutoff valve and is displayed on the center instrument panel. Fuel flow information is also provided to the FMS. (883 words)

NEW WORDS

dual	/ˈdjuːəl/	adj.	双的
turbofan	/ˈtəːbəufæn/	n.	涡扇
stage	/ˈsteidʒ/	n.	级
booster	/ˈbuːstə/	n.	增压器
shaft	/ˈʃɑːft/	n.	轴
turbine	/ˈtəːbain/	n.	涡轮
gearbox	/ˈgiəbɒks/	n.	变速箱；齿轮箱
overboost	/ˌəuvəˈbuːst/	n.	过度增压
imminent	/ˈiminənt/	adj.	迫近的
idle	/ˈaidl/	adj.&v.	闲置的；空转
vibration	/vaiˈbreiʃən/	n.	震动
windmilling	/ˈwindmiliŋ/	n.	风转；风磨
parameter	/pəˈræmitə/	n.	参数
illuminate	/iˈljuːmineit/	v.	闪烁，亮
profile	/ˈprəufail/	n.	剖面
hydro-mechanical	/ˈhaidrəumiˈkænikəl/	adj.	液压机械的
suppress	/səˈpres/	v.	压制
annunciator	/əˈnʌnʃieitə/	n.	信号器

contaminant	/kən'tæminənt/	*n.*	污染物
bypass	/'baipa:s/	*n.&v.*	旁通；绕开
saturated	/'sætʃəreitid/	*adj.*	饱和的
combustor	/kəm'bʌstə/	*n.*	燃烧室

EXPRESSIONS

starter motor   启动马达   reverse thrust   反推力

NOTES

MEC	main engine control	主发动机控制
EIS	Engine Instrument System	发动机仪表系统
PMC	power management control	动力管理控制

EXERCISES

Comprehension of Text A

I. Describe the given aviation terms from Text A in English.

1. auto-throttle   2. reverse thrust
3. dual rotor   4. turbofan
5. combustor

II. Answer the following questions after you have read Text A.

1. What type of powerplant does the B737-300 airplane have?

2. What is the main function of the main engine control?

3. How will the thrust be controlled?

4. When should we advance the thrust levers full forward?

5. What should the flight crew do if vibrations occur during descending?

6. When will the red warning light above EIS illuminate? And how can we cancel it?

7. What do oil pressure and oil temperature indications look like?

8. How do the MEC and PMC in the thrust control system operate?

9. How is fuel delivered to the combustor?

10. What is the function of the fuel filter?

Reading Aloud

III. Read the following paragraph aloud until you can say it in a natural way from your memory.

Flameout (熄火)

A flameout refers to the failure of a jet engine caused by the extinction of the flame in the combustion chamber. It can be caused by a number of factors, including fuel exhaustion, compressor stall, insufficient oxygen supply, foreign object damage such as birds, hail or even volcanic ash, severe weather conditions, and mechanical failure.

Flameouts occur most frequently when the engine is at an intermediate or low power setting (such as during the cruise and descent phases of flight). Most of the time, they are recovered from uneventfully. To recover from a flameout, the pilot should ensure the engine's fuel supply has been restored and then simply perform an engine restart as detailed in the aircraft's Flight Operations Manual.

Early jet engines were at relatively high risk of flameout. Fast acceleration or inappropriate throttle settings could impoverish (使……贫油) the fuel/air mixture causing a flameout. If this happened at low altitude, it would often lead to the total loss of the aircraft. However, modern jets are engineered to a higher degree of technical quality and are controlled by systems that constantly fine-tune their performance; as such flameouts are not such a risk as they were in the early days of jet-powered aviation. (201 words)

Vocabulary Practice

IV. Complete the following short passage by filling in the blanks with the words given in the box.

> checklist   overshooting   single   encountered   retract   cutoff
> malfunction   sufficient   reject   accelerate

Engine Failure during Takeoff

Although it is very unlikely to have an engine failure, pilots are trained to recognize engine failure and how to handle the situation. Below is a situation where an engine failure is 1_____during the takeoff roll for a typical B737.

Prior to takeoff, the takeoff thrust is set. It's normally around 40% initially. Then the TOGA button is pressed to make it 2_____to 90.2%-96.1%. If there's any 3_____before

80kts, the captain will 4_____the takeoff. After 80kts and before V₁ the captain will only reject for serious problems like engine failure or engine fire. After V₁ the takeoff must continue no matter what happens. This is to prevent the aircraft from 5_____the runway when the pilot tries to stop the plane.

V_1, V_R and V_2 are calculated before every flight. It depends on the runway length, aircraft weight, outside air temperature, and everything else. So if there's any malfunction after V_1, the takeoff will continue. Nothing will be done until 600 feet above ground level. Even if there is engine fire, no action will be taken until 600 feet. Take note that with a 6_____engine the climb will be slightly slower.

At 400feet, normally the captain will choose mode HDG SEL. And at 600 feet both pilots will determine the problem and if there's any recall 7_____to be done, now is the time to do it. If it's an engine failure, do nothing yet. At 800 feet, set flaps up speed around 210-220kts and 8_____flaps 5 to flaps 1 then flaps up. After the flaps are up, LVL CHG is selected and MAX CONT thrust is set with bank angle 25 degrees.

After all that, the Non-normal Checklist (NNC) begins. Thrust lever is closed and engine start lever is selected to 9_____, etc. After the Engine Failure NNC, the after Takeoff Checklist is accomplished.

The captain will inform the cabin while the co-pilot calls the control tower. The pilots will decide whether to continue or to come back and land. Normally, one engine is 10 to bring the plane safely to the original destination. (355 words)

Reading for More

V. Read the following passage and answer the given questions briefly.

Electrical System Malfunction

The loss of electrical power can deprive the pilot of numerous critical systems, and therefore should not be taken lightly even in day/VFR conditions. Most in-flight failures of the electrical system are located in the generator or alternator. Once the generator or alternator system goes off line, the electrical source in a typical light airplane is a battery. If a warning light or ammeter indicates the probability of an alternator or generator failure in an airplane with only one generating system, however, the pilot may have very little time available from the battery. The rating of the airplane battery provides a clue to how long it *may* last. With batteries, the higher the amperage (安培数) load, the less the usable total amperage. Thus a 25-amp (安培) hour battery could produce 5 amps per hour for 5 hours, but if the load were increased to 10 amps, it might last only 2 hours. A 40-amp load might discharge the battery fully in about 10 or 15 minutes. Much depends on the battery condition at the time of the system failure. If the battery has been in service for a few years, its power may be reduced substantially because of internal resistance (内电阻). Or if the system failure was not detected immediately, much of the stored energy may have already been used. It is essential, therefore, that the pilot immediately shed non-essential loads when the generating source fails. The pilot should then plan to land at the nearest suitable airport.

What constitutes an "emergency" load following a generating system failure cannot be predetermined, because the actual circumstances will always be somewhat different–for example, whether the flight is VFR or IFR, conducted in day or at night, in clouds or in the clear. Distance to nearest suitable airport can also be a factor.

The pilot should remember that the electrically powered (or electrically selected) landing gear and flaps will not function properly on the power left in a partially depleted battery. Landing gear and flap motors use up power at rates much greater than most other types of electrical equipment. The result of selecting these motors on a partially depleted (耗尽) battery may well result in an immediate total loss of electrical power. If the pilot should experience a complete in-flight loss of electrical power, the following steps should be taken:

• Shed all but the most necessary electrically driven equipment.

• Understand that any loss of electrical power is critical in a small airplane – notify ATC of the situation immediately. Request radar vectors for a landing at the nearest suitable airport.

• If landing gear or flaps are electrically controlled or operated, plan the arrival well ahead of time. Expect to make a no-flap landing, and anticipate a manual landing gear extension. (458 words)

Question 1　Why shouldn't electrical system malfunction be taken lightly?

Question 2　When the generating source fails, what is essential for the pilot to do?

Question 3　What should the pilot do after a complete in-flight loss of electrical power?

VI. In this section, there are two passages. After you have read the two passages, there are five questions followed by four choices marked A, B, C and D. Decide which one is the most appropriate answer.

Passage One

Engine Fire

An in-flight engine compartment fire is usually caused by a failure that allows a flammable substance such as fuel, oil or hydraulic fluid to come in contact with (接触) a hot surface. This may be caused by a mechanical failure of the engine itself, an engine-driven accessory, a defective induction or exhaust system, or a broken line. Engine compartment fires may also result from maintenance errors, such as improperly installed/fastened lines and/or fittings resulting in leaks.

Engine compartment fires can be indicated by smoke and/or flames coming from the engine cowling (整流罩) area. They can also be indicated by discoloration (退色) and/or melting of the engine cowling skin in cases where flames and/or smoke is not visible to the pilot. By the time a pilot becomes aware of an in-flight engine compartment fire, it usually is well developed. Unless the airplane manufacturer directs otherwise in the AFM/POH, the first step on discovering a fire should be to shut off the fuel supply to the engine by placing the mixture control in the idle cut off position and the fuel selector shutoff valve to the OFF position.

The ignition switch should be left ON in order to use up the fuel that remains in the fuel lines and components between the fuel selector/shutoff valve and the engine. This procedure may starve the engine compartment of fuel and cause the fire to *die* naturally. If the flames are out, no attempt should be made to restart the engine. If the engine compartment fire is oil-fed, as evidenced by thick black smoke, as opposed to a fuel-fed fire which produces bright orange flames, the pilot should consider stopping the propeller rotation by feathering(顺桨) or other means, such as (with constant-speed propellers) placing the pitch(桨距) control lever to the minimum r.p.m. position and raising the nose to reduce airspeed until the propeller stops rotating. This procedure will stop an engine-driven oil (or hydraulic) pump from continuing to pump the flammable fluid which is feeding the fire. (331 words)

1. An in-flight engine compartment fire is often caused by_____.
   A. some mechanical failure   B. maintenance errors
   C. improperly installed lines   D. flammable substances
2. A pilot usually notices an in-flight engine fire later than expected because of _____.
   A. his slow awareness of fire   B. unknown positions of fire
   C. uncertain fire indication   D. visible flames or smoke
3. What should pilots do first when discovering an engine fire?
   A. To shut off the fuel supply to the engine.
   B. To adjust the engine mixture control.
   C. To find out the exact location of engine fire.
   D. To switch the ignition switch to ON position.
4. After a fire is detected, the ignition switch should be left ON to_____.

A. use up the fuel left in lines　　　B. restart the engine in flight

C. shut off the engine fuel supply　　D. facilitate continuous ignition

5. The word *die* (Line 3, Para. 3) can be most appropriately replaced by_____.

A. snuff　　B. extinguish　　C. produce　　D. starve

Passage Two

Thrust Reverser

A thrust reverser is a device fitted in the engine exhaust system which effectively reverses the flow of the exhaust gases. The flow does not reverse through 180°; however, the final path of the exhaust gases is about 45° from straight ahead. This, together with the losses in the reverse flow paths, results in a net efficiency of about 50 percent. It will produce even less if the engine r.p.m. is less than maximum in reverse.

On most installations, reverse thrust is obtained with the thrust lever at idle, by pulling up the reverse lever to a detent (卡位). Doing so positions the reversing mechanisms for operation but leaves the engine at idle r.p.m. Further upward and backward movement of the reverse lever increases engine power. Reverse is cancelled by closing the reverse lever to the idle reverse position, then dropping it fully back to the forward idle position. This last movement operates the reverser back to the forward thrust position.

Reverse thrust is much more effective at high airplane speed than at low airplane speeds, for two reasons: first, the net amount of reverse thrust increases with speed; second, the power produced is higher at higher speeds because of the increased rate of doing work. In other words, the kinetic energy (动能) of the airplane is being destroyed at a higher rate at the higher speeds. To get maximum efficiency from reverse thrust, therefore, it should be used as soon as is prudent after touchdown.

When considering the proper time to apply reverse thrust after touchdown, the pilot should remember that some airplanes tend to pitch noseup when reverse is selected on landing and this effect, particularly when combined with the noseup pitch effect from the spoilers, can cause the airplane to leave the ground again momentarily. On these types, the airplane must be firmly on the ground with the nosewheel down, before reverse is selected. Other types of airplanes have no change in pitch, and reverse idle may be selected after the main gear is down and before the nosewheel is down. (342 words)

6. According to the passage, which of the following is not a factor of thrust reverser efficiency?

A. The direction of reversed flow.　　B. The path of reversed flow.

C. The engine speed.　　　　　　　D. Maximum reverse thrust.

7. More reverse thrust can be obtained _____.

   A. with the thrust levers in a more forward position

   B. with the thrust levers in a more backward position

   C. with the reverse levers in a more forward position

   D. with the reverse levers in a more backward position

8. Forward thrust can be available with the reverse levers in _____.

   A. reverse thrust position　　　　　B. idle reverse position

   C. idle thrust position　　　　　　D. forward thrust position

9. Reverse thrust is used right after touchdown so as to _____.

   A. make reverse thrust more efficient

   B. have the optimum aircraft attitude

   C. avoid any possible influences from weather

   D. increase kinetic energy during landing run

10. The nosewheel of some aircraft must firmly contact the ground before application of reverse thrust because the airplane _____.

   A. tends to pitch up on touchdown

   B. is less effective than tailwheels

   C. maintains a constant pitch attitude

   D. tends to be hard upon touchdown

VII. Read the following passage and retell it in your own words.

APU

   The Auxiliary Power Unit (APU) is a small jet engine located in the aircraft **tailcone**[1]. It allows the aircraft to be independent of external **pneumatic**[2] and electrical power supplies. The APU can provide **bleed air**[3] for starting engines and for air conditioning, and drives a **generator**[4] that provides electrical power. The **APU**[5] can be started and used on ground and in the air.

[1] 尾锥

[2] 气源

[3] 引气

[4] 发电机

[5] 辅助动力装置

   An APU start requires both the aircraft and APU batteries. The APU battery supplies power to the starter, the air inlet door, and the APU **control circuitry**[6]. The aircraft battery supplies power for fire protection, the APU fuel valve, and the DC fuel pump. It also supplies control circuitry power when the APU battery voltage drops because of heavy starter motor **load**[7]. Positioning the APU selector momentarily to

[6] 控制电路

[7] 载荷

START, begins the **start sequence** [8]. The APU fuel valve opens and at
the same time the APU inlet door starts to open. A fuel pump in the left
main tank also starts. A separate APU **transformer rectifier**[9] is
installed to power the APU starter when AC power is available. When
the inlet door is fully open, an electric starter engages, and the APU
accelerates to its normal operating speed. The starter duty cycle is a
maximum of 3 **consecutive** [10] starts or attempts within a 60 minute
period with a 5 minute cooling period between attempts. When the APU
RUN Light illuminates, the APU may be used to supply electrical power
and bleed air. Fuel for APU operation normally comes from the left main
tank. A DC fuel pump supplies the fuel when AC power is not available.
When AC power is available, the left forward AC fuel pump is signaled
to start regardless of its switch position. It then supplies fuel to the APU
and the DC pump remains off.

On the ground, placing the Battery Switch to OFF also causes an
APU shutdown but it should not be used for shutdown, because it could
result in the loss of APU **fire detection** [11]. In flight, the Battery Switch
does not affect the operation of the APU. An amber FAULT light on the
APU control panel illuminates whenever a fault or fire is sensed. In
addition, an EICAS advisory message APU FAULT is displayed, and the
APU shuts down immediately. Fault detection **circuitry**[12] is reset by
positioning the APU Selector to OFF. The FAULT light also comes on
when the APU fuel valve is not in the commanded position. Therefore,
during APU start and shutdown, the light illuminates momentarily. The
EICAS advisory APU FUEL VAL appears if the valve fails to reach the
signaled position. With the APU Selector positioned to OFF, both the
APU FAULT Light and the associated APU FAULT EICAS message are
inhibited. （464 words）

[8] 起动程序

[9] 变压整流器

[10] 连续的

[11] 火警探测

[12]电路，环路

Text B

Passage Rewriting

VIII. Read the following passage carefully and rewrite the passage under the given title with no less than 200 words.

In-flight Electrical Failure

_____

_____

The Electrical System

The function of the electrical system on a large transport aircraft is to generate, regulate, and distribute electrical power throughout the aircraft. Electrical power is used to operate aircraft flight instruments, essential systems, and passenger services. New generation aircraft are very dependent on electrical power because of the wide use of electronic flight instrument systems. Essential power is, just as the name implies, power that is essential to the safe operation of the aircraft. Power for passenger services is provided to light the cabin, to operate the entertainment system, and to prepare food. It is obvious that transport aircraft need a self-contained, dependable, and adequate power generating system.

The modern aircraft's electrical system is simply more automatic and easier to use. Generally speaking, there are two engine driven generators. The generators maintain a constant speed by a drive mechanism, known as an integrated drive generator (IDG).

Each generator supplies alternating current (AC) to its own bus: Generator 1 to AC bus 1, Generator 2 to AC bus 2. Each AC bus supplies its own transformer rectifier (TR).The TR converts alternating current into direct current (DC) to supply their associated DC buses, DC 1 and DC 2. DC bus 1 then feeds the DC BAT bus (DC BAT). The DC battery bus can charge the batteries or receive power from the batteries, as required. The electrical system also includes two essential buses. The first one is the AC ESS bus fed by AC bus 1 and the second one is the DC ESS bus fed by DC bus 1. This is the basic electrical system. There are some other components which supply the basic system. The electrical network can also be supplied by the APU generator. These three generators are all identical and any one of them can supply the entire aircraft electrical needs. On the ground, the aircraft can be supplied by an external power source. The electrical system is also fitted with an essential transformer rectifier. As a backup, there is a hydraulically driven emergency electrical generator (EMER GEN). The hydraulic power to drive the EMER GEN is provided by a ram air turbine (RAT) located in the belly fairing, which extends in case of severe electrical or hydraulic failures.

All the components talked about and information presented to the pilots are displayed on the ECAM page. Each component has a title to aid identification and each component can be monitored via its indications of voltage, current or load for the generators, frequency, and IDG temperature. The different connections are displayed via green lines.

The ELEC panel is located on the overhead panel. For emergency cases, there is an EMER ELEC PWR panel on the left side of the overhead panel. The battery voltage can be monitored either on the overhead panel or the ECAM page. Both main generators and the APU generator are controlled by their associated pushbutton switch.

The external power is also controlled by a pushbutton switch. The AC ESS FEED pushbutton switch enables the pilots to change the feed for the AC ESS bus from AC bus 1 to AC bus 2. The BUS TIE pushbutton switch enables the pilots to isolate one side of the system from the other. In case of failure, these pushbutton switches enable the flight crews to disconnect an IDG from its drive shaft. The EMER GEN TEST switch is used by maintenance only to test the emergency generator.

The initial indication of an electrical fire is usually the distinct odor of burning insulation. Once an electrical fire is detected, the pilot should attempt to identify the faulty circuit by checking circuit breakers, instruments, avionics, and lights. If the faulty circuit cannot be readily detected and isolated, and flight conditions permit, the battery master switch and alternator/generator switches should be turned off to remove the possible source of the fire. However, any materials which have been ignited may continue to burn.

If electrical power is absolutely essential for the flight, an attempt may be made to identify and isolate the faulty circuit by turning the electrical master switch OFF, turning all individual electrical switches OFF, turning the master switch back ON or selecting electrical switches that were ON before the fire indication one at a time, permitting a short time lapse after each switch is turned on to check for signs of odor, smoke, or sparks. This procedure, however, has the effect of recreating the original problem. The most prudent course of action is to land as soon as possible. (750 words)

NEW WORDS

generate	/'genəreit/	*v.*	发电
self-contained	/selfkən'teind/	*adj.*	自带的
battery	/'bætəri/	*n.*	电瓶
generator	/'genəreitə/	*n.*	发电机
pushbutton	/'puʃbʌtən/	*n.*	按钮
bus	/bʌs/	*n.*	汇流条
convert	/kən'və:t/	*v.*	转换
belly	/'beli/	*n.*	机腹
voltage	/'vəultidʒ/	*n.*	电压
current	/'kʌrənt/	*n.*	电流
insulation	/,insju'leiʃən/	*n.*	绝缘
alternator	/'ɔ:ltəneitə/	*n.*	交流机
lapse	/'læps/	*n.*	流逝（时间）
prudent	/'pru:dənt/	*adj.*	谨慎的

EXPRESSIONS

essential power	重要设备电源	circuit breakers	跳开关
transformer rectifier	变压整流器	BUS TIE	连接汇流条
master switch	总电门	drive shaft	驱动轴
belly fairing	机腹整流罩		

NOTES

IDG	integrated drive generator	综合驱动发电机
AC	alternating current	交流电
DC	direct current	直流电
RAT	ram air turbine	冲压空气涡轮
BAT	battery	电瓶
EMER ELEC PWR	emergency electrical power	应急电源
AC ESS bus	AC essential bus	交流主要设备汇流条

Lesson *8*

Pneumatics & Fuel

Preview

Different air sources supply air to the pneumatic system which serves to provide air conditioning, cabin pressurization and anti-icing. Some factors give rise to pressurization problems and consequent actions should be taken by pilots. Besides, fuel system is a part of the relevant knowledge that a pilot needs to acquire.

Text A

Warming-up Activities

Picture Description

Please describe the following picture and be prepared to answer some questions.

Relevant Questions

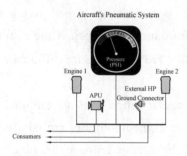

1. From the graph, what air sources does the aircraft pneumatic system have?
2. Which systems of an aircraft need engine bleed air?
3. What are the standard procedures for depressurization?
4. Is slow depressurization more dangerous than rapid depressurization? Why?

Cabin Pressurization

Pressurization of aircraft cabins above 3,000 meters (9,800 ft) generally avoids significant hypoxia, altitude sickness, decompression sickness and barotrauma. An oxygen system is retained but only for emergency use and only intended to allow time to descend to a safe altitude.

The pressure maintained within the cabin is referred to as the "cabin altitude". Cabin altitude is not normally maintained at ground level (0ft) pressure throughout the flight because doing so stresses the fuselage and uses more fuel. An aircraft planning to cruise at 40,000 ft (12,000 m) is programmed to rise gradually from take-off to around 8,000 ft (2,400 m) in cabin pressure altitude, and then to reduce gently to match the ambient air pressure of the destination. That destination may be significantly above sea level and this needs to be taken into account; for example, El Alto International Airport in La Paz, Bolivia is 4,061 meters (13,320 ft) above sea level.

Pressurization is achieved by the design of an airtight fuselage engineered to be pressurized, a source of compressed air and an environmental control system (ECS). The most

common source of compressed air for pressurization is bleed air extracted from the compressor stage of a gas turbine or turboprop propulsion engine, usually the second or third last compressor ring. By the time the cold outside air has reached this part of the compressor it has been heated to around 200 °C (392 °F) and is at a very high pressure. It is then expanded and cooled to a suitable temperature by passing it through a heat exchanger and air cycle machine (the packs system). There is no need to further heat or refrigerate the air. Typically, compressed air is bled from at least two propulsion engines, each system being fully redundant. Compressed air is also obtained from the Auxiliary Power Unit, if fitted, in the event of an emergency and for cabin air supply on the ground before the main engines are started.

All exhaust air is dumped to atmosphere via a valve, usually at the rear of the fuselage. This valve controls the cabin pressure and also acts as a safety relief. The pilot can alter the cabin pressure at will through this valve. Operational considerations typically require it to be set at 6,000 to 8,000 ft (1,800 to 2,400 m) – the latter being approximately 11 psi, giving a pressure differential between the cabin and the outside air of around 7.5 – 8 psi. If the cabin were maintained at sea level pressure while flown above 35,000 feet the pressure differential would exceed 9 psi limiting the structural life of the fuselage.

Cabin altitudes are maintained at up to 2,500 meters, so pressurization does not eliminate all physiological problems. Passengers with conditions such as a pneumothorax are advised not to fly until fully healed; pain may still be experienced in the ears and sinuses by people suffering from a cold or other infection.

The cabin of an aircraft is pressurized by using engine bleed air, which is conditioned and guided into the cabin. The amount and pressure of the conditioned air is regulated by a computer and outflow valves. Normally, pressurization problems are caused by a malfunctioning of the pressure outflow or the pressure regulating valves respectively, which allows the air to escape out of the pressurized cabin.

Pressurization problems can seriously impair the consciousness of the pilot. Therefore, a descent to altitudes less than 14 000 feet will be the immediate action. By doing this, the aircraft will descend through other levels and separation to other aircraft may be infringed. Pressurization problems can also cause an immediate danger to the aircraft, as the reduction of the cabin pressure will affect the crew and the passengers. The time for useful consciousness is dependent on the altitude flown, the size of the leak and the size of the fuselage. It can vary from 4 to 30 seconds.

In case of pressurization problems, the pilot will use an oxygen mask. In this case the headset is switched automatically to a cabin loudspeaker, with the pilot using the microphone in the mask which may reduce the readability to ATC. (688 words)

NEW WORDS

pneumatic	/nju:'mætik/	adj.	气源的
hypoxia	/hai'pɔksiə/	n.	缺氧
decompression	/di,kəm'preʃən/	n.	释压
barotrauma	/,bærə'trɔːmə/	n.	耳气压伤
oxygen	/'ɒksidʒən/	n.	氧气
retain	/ri'tein/	v.	保留
equivalent	/i'kwivələnt/	adj.	等量的
ambient	/'æmbiənt/	adj.	外界的
airtight	/'eətait/	adj.	气密的
refrigerate	/ri'fridʒərit/	v.	制冷
redundant	/ri'dʌndənt/	adj.	富余，多余的
duplicate	/'du:plikeit/	adj.	复制的
notoriety	/,nəutə'raiəti/	n.	恶名
pneumothorax	/,nju:mə'θɒræks/	n.	气胸
heal	/hi:l/	v.	治愈
sinus	/'sainəs/	n.	鼻窦
infection	/in'fekʃən/	n.	感染；传染
calibrate	/'kælibreit/	v.	校订；修正
regulate	/'regjuleit/	v.	调节
outflow	/'autfləu/	n.	外流
impair	/im'peə/	v.	损害
infringe	/in'frindʒ/	v.	侵犯
headset	/'hedset/	n.	耳机
loudspeaker	/'laud'spi:kə/	n.	扬声器
microphone	/'maikrəfəun/	n.	麦克风；话筒
readability	/,ri:də'biliti/	n.	清晰度
physiological	/,fiziə'lɒdʒikəl/	adj.	生理的

EXPRESSIONS

altitude sickness   恐高症   air cycle machine   空气循环机
pressure differential   压差   oxygen mask   氧气面罩

NOTES

psi   pound per square inch   磅/英寸^2

EXERCISES

Comprehension of Text A

I. Describe the given aviation terms from Text A in English.

1. cabin altitude
2. heat exchanger
3. outflow valve
4. separation

II. Answer the following questions after you have read Text A.

1. What are the benefits of pressurization of aircraft cabins?
2. Why are there oxygen systems in aircraft?
3. Why is the cabin altitude not normally maintained at ground level pressure throughout the flight?
4. How is pressurization achieved in the aircraft cabin?
5. What is the primary air source for aircraft pneumatic system?
6. What benefits does an APU bring to the pneumatic system?
7. How does the air in the cabin flow out of the airplane?
8. How is the pressure difference maintained between the cabin and the outside air?
9. What are the possible consequences of pressurization problems?
10. What actions should a pilot take if there is a pressurization problem?

Reading Aloud

III. Read the following paragraph aloud until you can say it in a natural way from your memory.

Smoke or Fire in the Cockpit

Smoke or fire in the cockpit is a serious emergency whose cause is generally a short circuit or other electrical malfunctions. It may run into heavy smoke and open fire in the cabin, possibly with serious consequences. Pilot visibility may be limited considerably for viewing cockpit instruments and outside so that aircraft may be totally reliant on ATC instructions.

In this situation, the pilots should be calm. It is always the first priority to keep control of the airplane. Then the flight crew, cabin members and passengers will put on oxygen masks in presence of smoke or poisonous gases. And the pilot should report the distress to ATC, start emergency descent and make immediate landing.

The air traffic controller should assist resolution of the situation by informing supervisor, asking for dangerous goods and number of persons on board (POB), informing landing aerodrome, clear RWY when ACFT is 25 NM final, keeping safety strip clear, and APP / RWY lighting system 100%. He should inform pilot about miles tracked until touchdown at next suitable aerodrome, availability of automatic approach low visibility procedure, details of landing aerodrome such as runway in use, length, surface, elevation, ILS and NAV – frequencies and weather information of landing aerodrome like wind, visibility, ceiling and QNH. (208 words)

Vocabulary Practice

IV. Complete the following short passage by filling in the blanks with the words given in the box.

> limiting   recommended   cabin altitude   safe altitude
>
> memory   idle   don   notify   target   initiate

Emergency Descent

The emergency descent is designed to bring the airplane down smoothly to a safe 1_____ in minimum time with the least possible passenger discomfort. It is intended as a specialized case to cover an uncontrollable loss of cabin pressurization.

If the descent is performed because of a rapid loss of cabin pressure, 2_____ oxygen masks and establish crew communication at the first indication of a loss of cabin pressurization. Verify that cabin pressure is uncontrollable, and if so, communicate with cabin crew to determine if obvious structural damage exists. Structural damage may require 3_____the descent speed.

All recall action items are to be accomplished by 4_____. The pilot not flying verifies that all recall items (记忆项) have been accomplished and call out any items not completed.

Emergency descent will be made with the landing gear up. Autopilot use is 5_____. In training, descent will be practiced both with and without autopilot.

To manually fly the maneuver, disconnect the autothrottle and retard the thrust levers to 6_____. Extend the speedbrakes, disconnect the autopilot and smoothly lower the nose down to 7_____ descent attitude. About 10 knots before reaching 8_____ speed (VMO/MMO) slowly reduce the pitch attitude to maintain target speed. Keep the airplane in trim at all times. If VMO/MMO is inadvertently exceeded, change pitch smoothly to decrease speed.

The pilot not flying will check minimum enroute altitude, 9_____ATC, obtain altimeter

setting(QNH) and call out altitudes approaching level-off (改平). Level off altitude should be chosen based on the oxygen system capacity or the lowest 10_____, whichever is higher. The lowest safe altitude is the Minimum Enroute Altitude(MEA), Minimum off-route Altitude(MORA), or any other altitude based on terrain clearance, navigation aid reception, or other appropriate criteria. The pilot not flying will call out 2,000 feet above and 1000 feet above level-off(改平) altitude.(302 words)

Reading for More

IV. Read the following passage and answer the given questions briefly.

Airframe Ice Control

Most wing de-icing and anti-icing equipment on aircraft fall within two categories–pneumatic boots (除冰带) and hot air. Pneumatic boots break ice from the wing by expanding, and the hot air system prevents ice formation by channeling (引导) hot air through the wing's leading edge. Light and medium reciprocating-engine aircraft usually use the pneumatic system while jet and turbine aircraft normally use the hot-air system.

A typical pneumatic wing de-icing system consists of inflation boots (充气带), a timer, pneumatic regulators, and engine-driven pressure and vacuum sources. Inflation boots are fabric-reinforced rubber sheets containing inflation tubes. The inflation boots are cemented to the leading edges of the wings, horizontal stabilizer, and vertical stabilizer. When you turn on the de-ice switch, all of the boots may operate simultaneously or alternately.

During normal operation, a slight vacuum pressure holds the boots in the deflated (泄气的) position. When you energize the system, positive pressure is applied to the inflation tubes. This pressure inflates the boots, separating the ice from the leading edge. The airflow over the airfoil carries the ice away.

Always follow the manufacturer's instructions for use of this type of equipment. As a general procedure, you normally operate the de-ice boots after one-fourth to one-half inch of ice has accumulated on the airfoil's leading edge. If you operate the boots with less ice accumulation, they tend to mold the ice to the new shape rather than breaking it from the airfoil surface. If this occurs, ice will accumulate on the contour (外形，轮廓) formed by the inflated boot and further operation of the de-ice system will have no effect.

A wing de-ice system may use the same pressure/vacuum source that powers some of the gyro instruments. On twin-engine aircraft, a pneumatic pump is usually located on each engine. In case of a single pump failure, the remaining pump supplies adequate pressure and vacuum to operate the gyro flight instruments and wing de-ice system.

Hot air anti-ice systems are commonly installed on turbojet and turboprop aircraft. This is because the jet engine has a ready source of hot air than can be used to heat the wing. In this system, hot air is bled form one of the stages of the compressor section of the engine and channeled to the aircraft wings. (373 words)

Question 1 How many categories do de-icing and anti-icing equipment on aircraft fall into?

Question 2 What is the operational procedure of the pneumatic wing deice system?

Question 3 Why is hot air anti-ice system commonly installed on turbojet and turboprop aircraft?

VI. In this section, there are two passages. After you have read the two passages, there are five questions followed by four choices marked A, B, C and D. Decide which one is the most appropriate answer.

Passage One

Cabin Fire

Cabin fires generally result from one of three sources: careless smoking on the part of the pilot and/or passengers; electrical system malfunctions; heating system malfunctions. A fire in the cabin presents the pilot with two immediate demands: *attacking* the fire, and getting the airplane safely on the ground as quickly as possible. A fire or smoke in the cabin should be controlled by identifying and shutting down the faulty system. In many cases, smoke may be removed from the cabin by opening the cabin air vents. This should be done only after the fire extinguisher (if available) is used. Then the cabin air  control can be opened to remove smoke and fumes. If smoke increases in intensity when the cabin air vents are opened, they should be immediately closed. This indicates a possible fire in the heating system, nose compartment baggage area (if so equipped), or that the increase in airflow is feeding the fire.

On pressurized airplanes, the pressurization air system will remove smoke from the cabin; however, if the smoke is intense, it may be necessary to either depressurize at altitude, if oxygen is available for all occupants, or execute an emergency descent.

In unpressurized single-engine and light twin-engine airplanes, the pilot can attempt to expel the smoke from the cabin by opening the weather windows. These windows should be closed immediately if the fire becomes more intense. If the smoke is severe, the passengers and crew should use oxygen masks if available, and the pilot should initiate an immediate descent. The pilot should also be aware that on some airplanes, lowering the landing gear and/or wing flaps can aggravate a cabin smoke problem. (276 words)

1. The word "attacking " (Line 6, Para. 1) most probably means_____.
   A. smoking   B. firing   C. controlling   D. identifying
2. What should a pilot immediately do when fire is found in the cabin?
   A. To find out the source of fire and put it out.
   B. To find out the source of fire and land quickly.
   C. To open the cabin air vents to remove the smoke.
   D. To use oxygen masks and descend immediately.
3. When you decide to open the cabin air vents to remove the smoke, you should first confirm _____.
   A. the cabin fire is under control   B. the faulty system is shut down
   C. the oxygen masks have been used   D. cabin altitude is in the safe range
4. On pressurized airplanes, if the smoke is intense and oxygen is available for all occupants, it may be necessary to _____.
   A. depressurize at altitude   B. execute an emergency descent
   C. make an emergency landing   D. depressurize or descend
5. The cabin smoke can get more serious when the pilot _____.
   A. retracts the wing flaps   B. opens the cabin air vents
   C. extends the wing flaps   D. opens cockpit windows

Passage Two

Door Opening in Flight

In most instances, the occurrence of an inadvertent (意外的) door opening is not of great concern to the safety of a flight, but rather, the pilot's reaction at the moment the incident happens. A door opening in flight may be accompanied by a sudden loud noise, sustained noise level and possible vibration or buffeting. If a pilot allows himself or herself to become distracted to the point where attention is focused on the open door rather than maintaining control of the airplane, loss of control may result, even though disruption (瓦解，扰乱) of airflow by the door is minimal.

In the event of an inadvertent door opening in flight or on takeoff, the pilot should adhere to the following. Concentrate on flying the airplane. Particularly in light single- and twin-engine airplanes; a cabin door that opens in flight seldom if ever compromises the airplane's ability to fly. There may be some handling effects such as roll and/or yaw, but in most instances these can be easily overcome. If the door opens after lift-off, do not rush to land. Climb to normal traffic pattern (起落航线) altitude, fly a normal traffic pattern, and make a normal landing. Do not release the seat belt and shoulder harness in an attempt to reach the door. Leave the door alone. Land as soon as practicable, and close the door once safely on the ground. Remember that most doors will not stay wide open. They will usually bang open, and then settle partly closed. A slip (侧滑) towards the door may cause it to open wider; a slip away from the door may push it closed.

Do not panic. Try to ignore the unfamiliar noise and vibration. Also, do not rush. Attempting to get the airplane on the ground as quickly as possible may result in steep turns at low altitude. Complete all items on the landing checklist. Remember that accidents are almost never caused by an open door. Rather, an open door accident is caused by the pilot's distraction or failure to maintain control of the airplane.

6. When a door is opened unexpectedly, it is _____.

    A. often a problem of flight safety

    B. more a matter of attention than safety

    C. an incident of noise abatement

    D. a problem of vibration or buffet

7. What is the pilot's first priority in case of a door opening in flight?

    A. Controlling the aircraft.

    B. Closing the door in time.

    C. Controlling the noise level.

    D. Focusing on the opened door.

8. With the cabin door open in flight, the author believes that _____.

    A. loss of control is about to result

    B. the opened door is not much a problem

    C. airplane's flying ability will change

    D. its handling effects need much concern

9. The pilot should close the door _____.

    A. right after the door opens in flight

    B. after normal operation is resumed

    C. once the aircraft lands safely

D. the seat belts are safely fastened

10. What is this passage mainly about?

A. The potential dangers of an inadvertent door opening.

B. Tips to follow in case of a door open in flight.

C. Possible causes for an inadvertent door opening.

D. How to prevent an inadvertent door opening.

VII. Read the following passage and retell it in your own words.

The Fuel System

The primary function of an aircraft fuel system is to supply clean fuel, free from vapor, at the proper pressures and flow rates to the engine under all operating conditions. The jet aircraft fuel system has fuel tanks located in the center fuselage area and the wings. The center tank is located in the fuselage. The inner tanks and the outer tanks are located in the wings. They are **replenish**ed [1] with fuel through a filler cap on top of the wing using a fuel nozzle and a **hose**[2] similar to the one to use to refuel an automobile. However, with airplanes, special attention should be taken not to allow the fuel nozzle **spout**[3] to project very far into the tank, since it can damage the tank.

[1] 重新装满, 补充
[2] 软管
[3] 管口, 喷口

Two fuel pumps are installed in the center tank. And two fuel pumps are installed in each inner tank. Two **transfer valves**[4] allow fuel to transfer from the outer tanks to the inner tanks. Each inner tank feeds its respective engine. Two engine low pressure valves are installed to cut off fuel to the engines. The centre tank also feeds the engines. A cross feed valve is fitted to connect or isolate the left and right sides. The fuel system also feeds the APU. A low pressure valve is fitted to cut off fuel to the APU. This represents the basic fuel system.

[4] 转换活门

Fuel is supplied from the aircraft fuel tanks through the necessary **filters boost pump**[5], and valves to the engine driven fuel pump. Fuel is delivered from the pump. The fuel control meters the fuel to the engine in proper quantities. Excess fuel is bypassed back to the pump.

[5] 过滤增压泵

Fuel tanks contain a vent that allows air pressure inside the tank to remain the same as that outside the tank. This prevents the formation of a vacuum that would prevent fuel from flowing out of the tank. The vents may be located in the **filler caps**[6], or the tank may be vented

[6] 加油口盖

through a small tube extending through the wing surface. The tanks also contain an overflow drain that prevents the **rupture**[7] of the tank due to   [7] 破裂 fuel expansion. The overflow drain may be combined with the fuel tank vent or it may have a separate opening. On hot days, it is not unusual to see a small amount of fuel coming from the overflow drain.

During the walk around, the flight crews must check several items which are part of the fuel system. These items are not major components but a check of their condition is very important. Fuel can be measured manually by the magnetic fuel level indicators located in each fuel tank. The flight crews are looking at a magnetic fuel level indicator in the right inner tank. It should be flush with the aircraft surface. Water drain valves should be checked that there is no water leaking from the valve. If there is a leak, call maintenance. (490 words)

Text B

Passage Rewriting

VIII. Read the following passage carefully and rewrite the passage under the given title with no less than 200 words.

Air Conditioning Failure

Air Conditioning System

The air conditioning system supplies conditioned bleed air and recirculated cabin air at a controlled temperature throughout the airplane. The system supplies conditioned air to the flight deck shoulder heaters. The system supplies ventilation for the passenger cabin, lavatories, and galleys. The system supplies ventilation for the flight deck crew rest compartment, lower crew rest compartment, and overhead crew rest compartments.

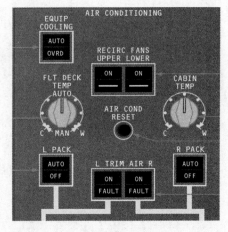

Pack control, zone temperature control, cabin air recirculation, fault detection, and overheat protection are all automatic. Backup system control modes operate automatically in the event of system failures.

Two identical air conditioning packs cool bleed air from the engines, APU, or high pressure air from a ground source. Bleed air is precooled before entering the pack. The packs are controlled by two identical pack controllers. If one controller fails, pack control switches automatically to the other controller. Pack output is increased automatically during high pack demand periods to compensate for a failed pack or recirculation fan, or is limited during high bleed air demand periods such as for gear retraction.

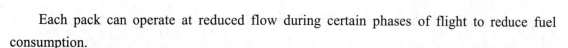

Each pack can operate at reduced flow during certain phases of flight to reduce fuel consumption.

Both air conditioning packs are normally selected to AUTO for ground operations. Fuel consumption is about the same for single pack and two pack operation, and single pack operation causes higher flight line noise levels.

Recirculation fans assist the packs to maintain a constant ventilation rate through the cabin. The fans draw cabin air through filters, then reintroduce the air into the conditioned air distribution system. The flight deck receives 100% fresh conditioned air from the left pack only, and is maintained at a slightly higher pressure than the passenger cabin to prevent smoke from entering the flight deck.

Air exhausted from the passenger cabin flows into the upper recirculation system or to the lower deck, where it is either exhausted overboard through outflow valves or drawn into the lower recirculation system. Air from the recirculation fans is mixed with pack air before entering the distribution ducts.

When one or more RECIRCULATION FAN switches is OFF, the EICAS memo message RECIRC FANS OFF is displayed. With the switches OFF, the packs operate at full flow and the cabin air exchange rate is decreased. Fuel consumption increases 0.7% for each fan switch turned OFF.

Hot trim air from the bleed air system is added to the pack conditioned air to control the temperature in each zone. Each trim air system supplies three zone supply ducts, with the left trim air system also supplying the flight deck. The cabin temperature controllers regulate the temperature by controlling the addition of hot trim air to the zone supply ducts through the trim air valves to meet the target temperature of each of the seven temperature zones. The CABIN TEMPERATURE selector sets a master reference temperature between 18 and 29 degrees C. The master reference temperature is increased or decreased automatically or manually to set target temperatures for each temperature zone.

For passenger comfort, the cabin temperature controllers compensate for temperature changes as cabin air humidity and passenger activity decrease during the flight. The target temperatures automatically increase slowly during the early part of cruise flight so the flight crew does not have to manually increase the master temperature. Target temperatures decrease slowly during descent until all automatically added corrections are removed.

The target temperatures of each passenger cabin temperature zone may be further modified plus or minus 6 degrees C, within the range of 18 to 29 degrees C, using the cabin management system.

The flight deck temperature selector sets the flight deck temperature to between 18 and 29 degrees C. The flight deck temperature is controlled manually by setting the flight deck

temperature selector to the manual mode. (621 words)

NEW WORDS

recirculate	/riˈsəːkjuleit/	v.	再循环
shoulder	/ˈʃəuldə/	n.	肩部
heater	/ˈhiːtə/	n.	加温器
ventilate	/ˈventileit/	v.	通风
galley	/ˈgæli/	n.	厨房
pack	/pæk/	n.	空调组件
precool	/priˈkuːl/	v.	预制冷
reintroduce	/ˌriːintrəˈdjuːs/	v.	再流入
overboard	/ˈəuvəbɔːd/	adv.	舱外
memo	/ˈmeməu/	n.	备忘，记忆
duct	/dʌkt/	n.	管道
humidity	/hjuːˈmiditi/	n.	湿度

EXPRESSIONS

trim air 调温空气

Lesson 9

Communication & Navigation

Preview

Various methods of navigation such as VOR, NDB, etc. and radio communication exist in current air navigation system. There are some airborne communication equipment like a transponder, and navigation systems including ADF, VOR, Inertial Navigation Systems, GPS for different IFR or VFR flight conditions.

Text A

Warming-up Activities

Picture Description

Please describe the following picture and be prepared to answer some questions.

Relevant Questions

1. What kinds of navaids do you know about?
2. What are functions of the common navigational equipment onboard an aircraft nowadays?
3. With advanced navigation systems onboard, are ground marks still important in modern navigation? Why?

Transponder

A transponder is an electronic device that produces a response when it receives a radio-frequency interrogation. In aviation, aircraft have transponders to assist in identifying them on radar and on other aircraft's collision avoidance systems. Air traffic control units use the term "squawk" when they are assigning an aircraft a transponder code, e.g. "Squawk 7421". Squawk or squawking thus can be said to mean "select transponder code" or "I have selected transponder code xxxx".

Depending on the type of interrogation, the transponder sends back a transponder code (Mode A) or altitude information (Mode C) to help air traffic controllers to identify the aircraft and to maintain separation. Another mode called Mode S (Mode Select) is designed to help avoid over-interrogation of the transponder (having many radars in busy areas) and to allow automatic collision avoidance. Mode S is mandatory in controlled airspace in many countries. Some countries have also required that all aircraft be equipped with Mode S, even in uncontrolled airspace.

Secondary Surveillance Radar is referred to as "secondary", to distinguish it from the "primary radar" that works by passively bouncing a radio signal off the skin of the aircraft. Primary radar works best with large all-metal aircraft, but not so well on small, composite aircraft. Its range is also limited by terrain and rain or snow and also detects unwanted objects such as automobiles, hills and trees. Furthermore not all primary radars can estimate the altitude of an aircraft. Secondary radar overcomes these limitations but it depends on a transponder in the aircraft to respond to interrogations from the ground station to make the aircraft more visible and to report the aircraft's altitude.

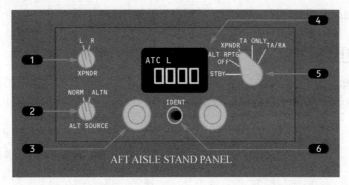

AFT AISLE STAND PANEL

Because primary radar generally gives bearing and range position information, but lacks altitude information, mode C and mode S transponders also report pressure altitude. Around busy airspace there is often a regulatory requirement that all aircraft be equipped with an altitude-reporting mode C or mode S transponders. Mode S transponders are compatible with transmitting the mode C signal, and have the capability to report in 25 foot increments. Without the pressure altitude reporting, the air traffic controller has no display of accurate altitude information, and must rely on the altitude reported by the pilot via radio. Similarly, the Traffic Collision Avoidance System installed on large aircraft as a last resort safety net needs the altitude information supplied by transponder signals.

All mode A, C, and S transponders include an "ident" button, which activates a special "thirteenth" bit on the mode A reply known as Ident. When radar equipment receives the Ident bit, it results in the aircraft's blip "blossoming" on the radar scope. This is often used by the controller to locate the aircraft amongst others by requesting the ident function from the pilot.

Ident can also be used in case of a reported or suspected radio failure to determine if the failure is only one way and whether the pilot can still transmit or receive but not both.

Transponder codes are four digit numbers transmitted by the transponder in an aircraft in response to a secondary surveillance radar interrogation signal to assist air traffic controllers in traffic separation. A discrete transponder code is assigned by air traffic controllers to uniquely identify an aircraft. This allows easy identity of the aircraft on radar.

Most codes above can be selected by aircraft if and when the situation requires or allows it, without permission from ATC. Other codes are generally assigned by ATC units. For IFR

flights, the squawk code is typically assigned as part of the departure clearance and stays the same throughout the flight. VFR flights, when in uncontrolled airspace, will "squawk VFR". Upon contact with an ATC unit, they will be told to squawk a certain unique code. When changing frequency, for instance because the VFR flight leaves controlled airspace or changes to another ATC unit, the VFR flight will be told to "squawk VFR" again.

In order to avoid confusion over assigned squawk codes, ATC units will typically be allocated blocks of squawk codes, not overlapping with the blocks of nearby ATC units, to assign at their discretion. (685 words)

NEW WORDS

transponder	/træns'pɒndə/	n.	应答机
interrogation	/inˌterə'geiʃən/	n.	询问
radar	/'reidə/	n.	雷达
squawk	/skwɔːk/	n. & v.	调定应答机
code	/kəud/	n.	编码
mandatory	/'mændətəri/	adj.	强制的
airspace	/'eəˌspeis/	n.	空域
surveillance	/sə'veləns/	n.	监控
bounce	/bauns/	v.	弹跳
composite	/kəm'pɒzit/	adj.	复合的
station	/steiʃən/	n.	电台
increment	/in'krimənt/	n.	增量
blip	/blip/	n.	（雷达）物体光点
transmit	/træns'mit/	v.	发射
discrete	/dis'kriːt/	adj.	分离的，分开的
convey	/kən'vei/	v.	传递
dial	/'daiəl/	n.	仪表盘
inclusive	/in'kluːsiv/	adj.	全部的
hijack	/'haiˌdʒæk/	n.	劫机
allocate	/'æləkeit/	v.	分配
overlap	/'əuvəlæp/	v.	重叠
discretion	/dis'kreʃən/	n.	谨慎

EXPRESSIONS

Secondary Surveillance Radar　二次雷达　　radar scope　雷达显示器
primary radar　一次雷达　　　　　　　be compatible with　与……相容

EXERCISES

Comprehension of Text A

I. Describe the given aviation terms from Text A in English.

1. aircraft's collision avoidance system
2. transponder code
3. Secondary Surveillance Radar
4. emergency code
5. TCAS

II. Answer the following questions after you have read Text A.

1. What is a transponder according to the passage?
2. What are the functions of transponders in aviation?
3. What does the term "squawk" mean?
4. How many types of transponder codes are mentioned in the text? What are they?
5. What are the major differences between Secondary Surveillance Radar and Primary Radar?
6. In what conditions can we use the "Ident" button on transponders?
7. How is the transponder ident activation displayed on ATC radars?
8. How do air traffic controllers assign transponder codes to aircraft?
9. What should a pilot do when he leaves a controlled airspace in VFR flight?
10. Why are blocks of squawk codes allocated between nearby ATC units?

Reading Aloud

III. Read the following paragraph aloud until you can say it in a natural way from your memory.

Communication Failure

Communication failure describes an aircraft without a radio (NORDO), even among pilots and others who are not air traffic controllers. While sometimes used to denote small general aviation aircraft that are not equipped with one, the term is more commonly applied to aircraft that have experienced a radio failure in midair. This may constitute an emergency, as determined by the pilot. Aircraft equipped with a transponder should indicate a NORDO situation by setting the appropriate transponder code: 7600. NORDO aircraft declaring an emergency are given priority over other aircraft if a more serious emergency does not occur on

another aircraft.

If the radio failure occurs in VFR conditions in an area where radio communication is required, the pilot is expected to continue under VFR and land when feasible. In IFR conditions, the pilot is expected to follow the last instructions given.

Air traffic control may re-establish communications with NORDO aircraft by using emergency frequencies, voice features of NAVAIDs, or aviation light signals. In the event of one-way communications (i.e. aircraft can receive only), the controller may request the aircraft make identifying turns, flash their navigational lights, transmit codes or IDENT signals on the transponder, rock their wings, etc, to acknowledge clearances or instructions. (202 words)

Vocabulary Practice

IV. Complete the following short passage by filling in the blanks with the words given in the box.

> flying   course   instrument   followed   transmitter
> relative   needle   frequency   indicate   select

Using VOR

For VOR radio navigation, there are two components required: the ground 1_____ and the airplane receiving equipment. The ground transmitter is located at a specific position on the ground and transmits on an assigned 2_____. The airplane equipment includes a receiver with a tuning device and a VOR or omninavigation 3_____. The navigation instrument consists of an omnibearing selector (OBS) sometimes referred to as the course selector, a 4 deviation indicator needle, and a TO-FROM indicator. The course selector is an azimuth(方位) dial that can be rotated to 5_____ a desired radial or to determine the radial over which the airplane is 6_____. In addition, the magnetic course "TO" or "FROM" the station can be determined.

When the course selector is rotated, it moves the course deviation indicator (CDI) or needle to indicate the position of the radial 7_____ to the airplane. If the course selector is rotated until the deviation needle is centered, the radial (magnetic course "FROM" the station) or its reciprocal (magnetic course "TO" the station) can be determined. The course deviation 8_____ will also move to the right or left if the airplane is flown or drifting away from the radial which is set in the course selector.

By centering the needle, the course selector will 9_____ either the course "FROM" the

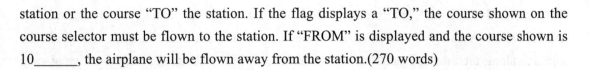
station or the course "TO" the station. If the flag displays a "TO," the course shown on the course selector must be flown to the station. If "FROM" is displayed and the course shown is 10_____, the airplane will be flown away from the station.(270 words)

Reading for More

V. Read the following passage and answer the given questions briefly.

Pilotage and Dead Reckoning (地标领航与推测领航)

Pilotage is navigation by reference to landmarks or checkpoints. It is a method of navigation that can be used on any course that has adequate checkpoints, but it is more commonly used in conjunction with dead reckoning and VFR radio navigation.

The checkpoints selected should be prominent features common to the area of the flight. Choose checkpoints that can be readily identified by other features such as roads, rivers, railroad tracks, lakes, and power lines. If possible, select features that will make useful boundaries or brackets on each side of the course, such as highways, rivers, railroads, and mountains. A pilot can keep from drifting too far off course by referring to and not crossing the selected brackets. Never place complete reliance on any single checkpoint. Choose ample checkpoints. If one is missed, look for the next one while maintaining the heading. When determining position from checkpoints, remember that the scale of a sectional chart is 1 inch = 8 statute miles or 6.86 nautical miles. For example, if a checkpoint selected was approximately one-half inch from the course line on the chart, it is 4 statue miles or 3.43 nautical miles from the course on the ground. In the more congested areas, some of the smaller features are not included on the chart. If confused, hold the heading. If a turn is made away from the heading, it will be easy to become lost.

Roads shown on the chart are primarily the well-traveled roads or those most apparent when viewed from the air. New roads and structures are constantly being built, and may not be shown on the chart until the next chart is issued. Some structures, such as antennas may be difficult to see. Sometimes TV antennas are grouped together in an area near a town. They are supported by almost invisible guy wires. Never approach an area of antennas less than 500 feet above the tallest one. Most of the taller structures are marked with strobe lights to make them more visible to a pilot. However, some weather conditions or background lighting may make them difficult to see. Aeronautical charts display the best information available at the time of printing, but a pilot should be cautious for new structures or changes that have occurred since the chart was printed.

Dead reckoning is navigation solely by means of computations based on time, airspeed, distance, and direction. The products derived from these variables, when adjusted by windspeed and velocity, are heading and groundspeed. The predicted heading will guide the airplane along the intended path and the groundspeed will establish the time to arrive at each checkpoint and the destination. Except for flights over water, dead reckoning is usually used with pilotage for cross-country flying. The heading and groundspeed as calculated is constantly monitored and corrected by pilotage as observed from checkpoints. (471 words)

Question 1　What features of checkpoints are mentioned in the passage?

Question 2　Why is the way of ensuring adequate checkpoints recommended by the author?

Question 3　What factors are taken into account for the computation in dead reckoning?

VI. In this section, there are two passages. After you have read the two passages, there are five questions followed by four choices marked A, B, C and D. Decide which one is the most appropriate answer.

Passage One

Getting Lost

Getting lost in an airplane is a potentially dangerous situation especially when low on fuel. If a pilot becomes lost, there are some good common sense procedures to follow. If a town or city cannot be seen, the first thing to do is climb, being mindful of traffic and weather conditions. An increase in altitude increases radio and navigation reception range, and also increases radar coverage. If flying near a town or city, it might be possible to read the name of the town on a water tower.

If the airplane has a navigational radio, such as a VOR or ADF receiver, it can be possible to determine position by plotting an azimuth (方位) from two or more navigational facilities. If GPS is installed, or a pilot has a portable aviation GPS on board, it can be used to determine the position and the location of the nearest airport.

Communicate with any available facility using frequencies shown on the sectional chart. If contact is made with a controller, radar vectors may be offered. Other facilities may offer direction finding (DF) assistance. To use this procedure, the controller will request the pilot to hold down the transmit button for a few seconds and then release it. The controller may ask the pilot to change directions a few times and repeat the transmit procedure. This gives the controller enough information to plot the airplane position and then give vectors to a suitable landing site. If the situation becomes threatening, transmit the situation on the emergency frequency 121.5 MHz and set the transponder to 7700. Most facilities, and even airliners, monitor the emergency frequency. (270 words)

1. It is advisable to climb in case of getting lost because _____.

   A. higher altitude means better reception of navigational signals

   B. the pilot can have more fields of vision at higher altitudes

   C. aircraft are more subject to conflicting traffic at lower level

   D. airborne radar equipment tends to operate better at lower level

2. When a pilot encounters disorientation, he should be aware of _____.

   A. aircraft performance   B. radar coverage

   C. water tower   D. current weather

3. Which of the following is an implied means of determining position?

   A. ADF receiver.   B. GPS receiver.   C. VOR receiver.   D. Ground marks

4. Frequencies are used to contact ATC to help a pilot _____.

   A. adjust navigational equipments   B. make sure of his present position

   C. reduce possible communication   D. reduce workload in the cockpit

5. The pilots may be required to make some maneuvers _____.

   A. to increase aircraft altitude   B. to increase fuel economy

   C. to establish ATC radar vectors   D. to monitor navigational equipments

Passage Two

Automatic Direction Finder

Many general aviation-type airplanes are equipped with automatic direction finder (ADF) radio receiving equipment. To navigate using the ADF, the pilot tunes the receiving equipment to a ground station known as a Non-Directional Beacon (NDB). The NDB stations normally operate in a low or medium frequency band of 200 to 415 kHz. The frequencies are readily available on aeronautical charts or in the Airport/Facility Directory.

All radio beacons except compass locators transmit a continuous three-letter identification code except during voice transmissions. A compass locator, which is associated with an Instrument Landing System, transmits a two-letter identification. Standard broadcast stations can also be used in conjunction with ADF. Positive identification of all radio stations is extremely important and this is particularly true when using standard broadcast stations for navigation. Nondirectional radio beacons have one advantage over the VOR. This advantage is that low or medium frequencies are not affected by line-of-sight. The signals follow the curvature of the Earth; therefore, if the airplane is within the range of the station, the signals can be received regardless of altitude.

Keep in mind that the needle of fixed azimuth points to the station in relation to the nose of the airplane. If the needle is deflected 30° to the left or a relative bearing of 330°, this means that the station is located 30° left. If the airplane is turned left 30°, the needle will move to the right 30° and indicate a relative bearing of 0° or the airplane will be pointing toward the station. If the pilot continues flight toward the station keeping the needle on 0°, the procedure is called homing to the station. If a crosswind exists, the ADF needle will continue to drift away from zero. To keep the needle on zero, the airplane must be turned slightly resulting in a curved flight path to the station. Homing to the station is a common procedure, but results in drifting downwind, thus lengthening the distance to the station. Tracking to the station requires correcting for wind drift and results in maintaining flight along a straight track or bearing to the station. When the wind drift (偏流) correction is established, the ADF needle will indicate the amount of correction to the right or left. For instance, if the magnetic bearing to the station is 340°, a correction for a left crosswind would result in a magnetic heading of 330°, and the ADF needle would indicate 10° to the right or a relative bearing of 010°.

When tracking away from the station, wind corrections are made similar to tracking to the station, but the ADF needle points toward the tail of the airplane or the 180° position on the azimuth dial (方位刻度). Attempting to keep the ADF needle on the 180° position during winds results in the airplane flying a curved flight leading further and further from the desired track. To correct for wind when tracking outbound, correction should be made in the direction opposite of that in which the needle is pointing. (514 words)

6. We can learn from the passage that ADF _____.

    A. functions in a higher frequency band

    B. serves as an onboard equipment

    C. is independent of ground stations

    D. is solely applied in general aviation

7. A compass locator is unique because it _____.

    A. can not be found in aeronautical charts

    B. has the only function of voice transmission

    C. is used in precision approach system

    D. can not be found in aeronautical charts

8. NDB is better than VOR because _____.

    A. its signals can be affected by the curvature of the earth

    B. it can be used in conjunction with broadcast stations

    C. medium frequencies are heavily affected by line-of-sight

    D. its signals can be received regardless of altitude

9. What is said in the passage about homing to the station?

    A. The needle of fixed azimuth remains stable.

B. Crosswind has some effects on flight paths.

C. The needle will not always points to the station.

D. Drift correction has no indications to pilots.

10. When flying from the station, _____.

A. the ADF needle would indicate 10° to the right

B. the ADF needle points toward the rear of the airplane

C. maintaining needle position leads to a curved flight

D. wind drift should be corrected as when flying inbound

VII. Read the following passage and retell it in your own words.

Inertial Navigation System

An inertial navigation system includes at least a computer and a module containing **accelerometers**[1], gyroscopes, or other motion-sensing devices. The INS is initially provided with its position and velocity from another source (a **human operator**[2], a GPS satellite receiver, etc.), and thereafter computes its own updated position and velocity by integrating information received from the motion sensors. The advantage of an INS is that it requires no external references in order to determine its position, **orientation**[3], or velocity once it has been initialized.

[1] 加速计

[2] 操作人员

[3] 方向

An INS can detect a change in its geographic position, a change in its velocity, and a change in its orientation (rotation about an axis). It does this by measuring the linear and **angular**[4] accelerations applied to the system. Since it requires no external reference (after **initialization**[5]), it is immune to jamming and deception.

[4] 角度的

[5] 初始化

Gyroscopes measure the angular velocity of the system in the inertial reference frame. By using the original orientation of the system in the inertial reference frame as the initial condition and integrating the angular velocity, the system's current orientation is known at all times. This can be thought of as the ability of a **blindfolded**[6] passenger in a car to feel the car turn left and right or **tilt**[7] up and down as the car ascends or descends hills. Based on this information alone, he knows what direction the car is facing but not how fast or slow it is moving, or whether it is sliding sideways.

[6] 盲目的

[7] 倾斜

Accelerometers measure the **linear**[8] acceleration of the system in the inertial reference frame, but in directions that can only be measured relative to the moving system. The accelerometers are fixed to and rotate

[8] 线性的

with the system, but are not aware of their own orientation.

However, by tracking both the current angular velocity of the system and the current linear acceleration of the system measured relative to the moving system, it is possible to determine the linear acceleration of the system in the inertial reference frame. Performing integration on the inertial accelerations (using the original velocity as the initial conditions) using the correct kinematic **equations**[9] yields the inertial velocities of the system, and integration again (using the original position as the initial condition) yields the inertial position. In our example, if the blindfolded passenger knew how the car was pointed and what its velocity was before he was blindfolded, and if he is able to keep track of both how the car has turned and how it has accelerated and **decelerated**[10] since, he can accurately know the current orientation, position, and velocity of the car at any time. (446 words)

[9] 方程

[10] 减速

Text B

Passage Rewriting

VIII. Read the following passage carefully and rewrite the passage under the given title with no less than 200 words.

Navigation in VFR

Air Navigation

The first step in navigation is deciding where one wishes to go. A private pilot planning a flight under VFR will usually use an aeronautical chart of the area which is published specifically for the use of pilots. This map will depict controlled airspace, radio navigation aids and airfields prominently, as well as hazards to flying such as mountains, tall radio masts, etc. It also includes sufficient ground detail – towns, roads, wooded areas – to aid visual navigation.

The pilot will choose a route, taking care to avoid controlled airspace that is not permitted for the flight, restricted areas, danger areas and so on. The chosen route is plotted on the map, and the lines drawn are called the track. The aim of all subsequent navigation is to follow the chosen track as accurately as possible. Occasionally, the pilot may elect on one leg to follow a clearly visible feature on the ground such as a railway track, river, highway, or coast.

When an aircraft is in flight, it is moving relative to the body of air through which it is flying; therefore, maintaining an accurate ground track is not as easy as it might appear, unless there is no wind at all. The pilot must adjust heading to compensate for the wind, in order to follow the ground track. Initially the pilot will calculate headings to fly for each leg of the trip prior to departure, using the forecast wind directions and speeds supplied by the

meteorological authorities for the purpose. These figures are generally accurate and updated several times per day, but the unpredictable nature of the weather means that the pilot must be prepared to make further adjustments in flight.

The primary instrument of navigation is the magnetic compass. The needle or card aligns itself to magnetic north, which does not coincide with true north, so the pilot must also take magnetic variation into consideration. The local variation is also shown on the flight map. Once the pilot has calculated the actual headings required, the next step is to calculate the flight time for each leg. This is necessary to perform accurate dead reckoning. The pilot also needs to take into account the slower initial airspeed during climb to calculate the time to top of climb. It is also helpful to calculate the top of descent, or the point at which the pilot would plan to commence the descent for landing.

The flight time will depend on both the desired cruising speed of the aircraft, and the wind – a tailwind will shorten flight times, a headwind will increase them.

The point of no return is the point on a flight at which a plane has just enough fuel, plus any mandatory reserve, to return to the airfield from which it departed. Beyond this point, the plane must proceed to some other destination. Alternatively, with respect to a large region without airfields, it can mean the point before which it is closer to turn around and after which it is closer to continue. Similarly, the Equal time point, referred to as the ETP, is the point in the flight where it would take the same time to continue flying straight, or track back to the departure aerodrome. The ETP is not dependant on fuel, but on wind, giving a change in ground speed out from, and back to the departure aerodrome. In Nil wind conditions, the ETP is located halfway between the two aerodromes, but in reality it is shifted depending on the wind speed and direction.

Commercial aircraft are not allowed to operate along a route that is out of range of a suitable place to land if an emergency such as an engine failure occurs. The ETP calculations serve as a planning strategy, so flight crew always have an option in an emergency event, allowing a safe diversion to their chosen alternate.

The final stage is to note which areas the route will pass through or over, and to make a note of all of the things to be done–ATC units, the appropriate frequencies, visual reporting points, and so on. It is also important to note which pressure setting regions will be entered, so that the pilot can ask for the QNH (air pressure) of those regions. Finally, the pilot should have in mind some alternative plans in case the route cannot be flown for some reason like unexpected weather conditions. At times the pilot may be required to file a flight plan for an alternate destination and to carry adequate fuel for this. The more work a pilot can do on the ground prior to departure, the easier it will be in the air. (782 words)

NEW WORDS

aeronautical	/ˌeərəˈnɔtikəl/	*adj.*	航空的
airfield	/ˈeəfiːld/	*n.*	机场
mast	/mæst/	*n.*	杆
plot	/plɒt/	*v.*	标出
aerodrome	/ˈeərədrəum/	*n.*	机场
meteorological	/ˌmiːtiərəˈlɒdʒikəl/	*adj.*	气象的
leg	/leg/	*n.*	航段
feature	/ˈfitʃə/	*n.*	装置
variation	/ˌveəriˈeiʃən/	*n.*	变化，偏差
forecast	/ˌfɔːkɑːst/	*v.*	预报
compass	/ˈkʌmpəs/	*n.*	罗盘
coincide	/ˌkəuinˈsaid/	*v.*	巧合；一致

EXPRESSIONS

aeronautical chart 航空地图	magnetic compass 磁罗盘
navigation aids 导航设备	visual navigation 目视导航
restricted area 限制区；限航区	magnetic variation 磁偏角

Unit Three
Procedures

Lesson 10

On the Ground

Preview

Preflight is the first important part of any flight. A series of tasks such as preflight briefing, weather check, preflight inspection, etc. have to be completed to make sure that the aircraft is airworthy and that the flight and cabin crew are completely prepared in every aspect. Text A describes the weight and balance which have a great influence on aircraft performance. Text B helps us have a better understanding of evacuation slides.

Text A

Warming-up Activities

Picture Description

Please describe the following picture and be prepared to answer some questions.

Relevant Questions

1. Do you know anything about ground incidents and ground accidents?
2. In what conditions are ground events more likely to occur?
3. How will you make a normal preflight inspection?
4. Will company procedures effectively help reduce ground incidents? Why?

Weight and Balance

There are many factors that lead to efficient and safe operation of aircraft. Among these vital factors is proper weight and balance control. The weight and balance system commonly employed among aircraft consists of three equally important elements: the weighing of the aircraft, the maintaining of the weight and balance records, and the proper loading of the aircraft. An inaccuracy in any one of these elements affects the purpose of the whole system. The final loading calculations will be meaningless if either the aircraft has been improperly weighed or the records contain an error.

Improper loading cuts down the efficiency of an aircraft from the standpoint of altitude,

142

maneuverability, rate of climb, and speed. It may even be the cause of failure to complete the flight, or failure to start the flight. Because of abnormal stresses placed upon the structure of an improperly loaded aircraft, or because of changed flying characteristics of the aircraft, loss of life and destruction of valuable equipment may result.

The responsibility for proper weight and balance control begins with the engineers and designers, and extends to the aircraft mechanics that maintain the aircraft and the pilots who operate them.

The designers of an aircraft have set the maximum weight, based on the amount of lift the wings or rotors can provide under the operation conditions for which the aircraft is designed. The structural strength of the aircraft also limits the maximum weight the aircraft can safely carry. The ideal location of the center of gravity (CG) was very carefully determined by the designers, and the maximum deviation allowed from this specific location has been calculated.

When an aircraft is designed, it is made as light as the required structural strength will allow, and the wings are designed to support the maximum allowable weight. When the weight of an aircraft is increased, the wings must produce additional lift and the structure must support not only the additional static loads, but also the dynamic loads imposed by flight maneuvers.

The pilot in command of the aircraft has the responsibility on every flight to know the maximum allowable weight of the aircraft and its CG limits. This allows the pilot to determine on the preflight inspection that the aircraft is loaded in such a way that the CG is within the allowable limits.

Severe uncoordinated maneuvers or flight into turbulence can impose dynamic loads on the structure great enough to cause failure. The structure of a normal category airplane must be strong enough to sustain a load factor of 3.8 times its weight. An aircraft operated in the utility category must sustain a load factor of 4.4, and acrobatic category aircraft must be strong enough to withstand 6.0 times their weight.

The lift produced by a wing is determined by its airfoil shape, angle of attack, speed through the air, and the air density. When an aircraft takes off from an airport with a high density altitude, it must accelerate to a speed faster than would be required at sea level to produce enough lift to allow takeoff; therefore, a longer takeoff run is necessary. The distance needed may be longer than the available runway. When operating from a high-density altitude airport, the Pilot Operating Handbook (POH) or Airplane Flight Manual (AFM) must be consulted to determine the maximum weight allowed for the aircraft under the conditions of altitude, temperature, wind, and runway conditions.

Most modern aircraft are so designed that if all seats are occupied, all baggage allowed by the baggage compartment is carried, and all of the fuel tanks are full, the aircraft will be grossly overloaded. This type of design requires the pilot to give great consideration to the requirements of the trip. If maximum range is required, occupants or baggage must be left behind, or if the maximum load must be carried, the range, dictated by the amount of fuel on board, must be reduced.

There are some problems caused by overloading an aircraft. The aircraft will need a higher takeoff speed, which results in a longer takeoff run. Both the rate and angle of climb will be reduced, the service ceiling will be lowered. The cruising speed will be reduced, the cruising range shortened and maneuverability decreased. A longer landing roll will be required because the landing speed will be higher. Excessive loads will be imposed on the structure, especially the landing gear.

The POH or AFM includes tables or charts that give the pilot an indication of the performance expected for any weight. An important part of careful preflight planning includes a check of these charts to determine the aircraft is loaded so the proposed flight can be safely made. (778 words)

NEW WORDS

inaccuracy	/inˈækjurəsi/	*n.*	不准确
nullify	/ˈnʌlifai/	*v.*	使无效
stress	/stres/	*n.*	应力，压力
structure	/ˈstrʌktʃə/	*n.*	结构
destruction	/diˈstrʌkʃən/	*n.*	破坏
operator	/ˈɒpəreitə/	*n.*	运营人
amateur	/ˈæmətəː/	*adj.*	业余的
alteration	/ˌɔːltəˈreiʃən/	*n.*	更改
diligence	/ˈdilidʒəns/	*n.*	勤奋
acrobatic	/ˌækrəˈbætik/	*adj.*	表演的
grossly	/ˈɡrəusli/	*adv.*	过多地

EXPRESSIONS

load factor	负载因素	structural strength	结构强度
Airplane Flight Manual	飞机飞行手册	uncoordinated maneuvers	不协调机动
dynamic load	动态负载	preflight planning	飞行前计划

EXERCISES

Comprehension of Text A

I. Describe the given aviation terms from Text A in English.

1. load factor
2. flying characteristics
3. structural strength
4. center of gravity
5. maximum allowable weight
6. uncoordinated maneuvers

II. Answer the following questions after you have read Text A.

1. What elements does the weight and balance system of an aircraft have?
2. What consequences can improper loading result in?
3. Who is responsible for proper weight and balance control of an aircraft?
4. What factors may influence the maximum weight of an aircraft?
5. How is the center of gravity of an aircraft determined?
6. What is the responsibility of the pilot in command in preflight inspection in terms of weight and balance control?
7. Why is weight factor very important?
8. In designing an aircraft, how is weight related to structural strength?
9. When an aircraft takes off from an airport with a high density altitude, what factors should a pilot take into consideration?
10. What problems are caused by overloading an aircraft?

Reading Aloud

III. Read the following paragraph aloud until you can say it in a natural way from your memory.

Contaminated Runways（污染跑道）

In civil aviation, a contaminated runway is one that is covered in a relatively deep layer of water, slush, loose snow, ice or compacted snow. The direct effects on aircraft performance of such contaminants arise due to the additional drag of the contaminants on the tyres and the reduced braking friction available.

The consequences are mixed. Takeoff and landing distances in pilot's operating handbooks are based on paved, dry, and level runway conditions. A contaminated runway would considerably increase the overall takeoff roll. During the take-off ground run the extra drag on the wheels reduces the aircraft's ability to accelerate, and longer runway length is required to accelerate to takeoff speed. During landing or aborted takeoff, the reduced braking friction and increased drag on the tyres act in opposition to one another. In this case, more effective use of thrust reverser, brakes, speedbrakes and rudder pedals is required in order to avoid overrun. In addition, the presence of such contaminants can also affect severely the ground-handling capability of the aircraft, particularly in cross-wind conditions.

To avoid running out of runway, look up your pilot's operating manual for distances required to make a takeoff or landing under the conditions that exist at the time of the operation. (205 words)

Vocabulary Practice

IV. Complete the following short passage by filling in the blanks with the words given in the box.

> alert landing detected blinding cockpit
> propeller available airflow battery place

Starting, Taxiing and Runup

After the pilot is seated in the 1_____and prior to starting the engine, all items and materials to be used on the flight should be arranged in such a manner that they will be readily 2_____and convenient to use.

Extra caution should be taken at night to assure the 3_____area is clear. Turning the rotating beacon ON, or flashing the airplane position lights will serve to 4_____persons nearby to remain clear of the propeller. To avoid excessive drain of electrical current from the 5_____, it is recommended that unnecessary electrical equipment be turned OFF until after the engine has been started.

After starting and before taxiing, the taxi or 6_____light should be turned ON. Continuous use of the landing light with r.p.m. power settings normally used for taxiing may 7_____an excessive drain on the airplane's electrical system. Also, overheating of the landing light could become a problem because of inadequate 8_____ to carry the heat away. Landing lights should be used as necessary while taxiing. When using landing lights, consideration should be given to not 9_____other pilots. Taxi slowly, particularly in congested areas. If taxi lines are painted on the ramp or taxiway, these lines should be

followed to ensure a proper path along the route.

The before-takeoff and runup (试车) should be performed using the checklist. During the day, forward movement of the airplane can be 10_____ easily. At night, the airplane could creep forward without being noticed unless the pilot is alert for this possibility. Hold or lock the brakes during the runup and be alert for any forward movement. (269 words)

Reading for More

V. Read the following passage and answer the given questions briefly.

Pushback（推出）

In aviation, pushback is an airport procedure during which an aircraft is pushed backwards away from an airport gate by external power. Pushbacks are carried out by pushback tractors or tugs.

Although many aircraft can also move backwards on the ground using reverse thrust, the resulting jet blast (喷气机喷气) may cause damage to the terminal building or equipment. Engines close to the ground may also blow sand and debris(碎片) forward and then suck it in to the engine, causing damage to the engine. A pushback using a tractor is therefore the preferred method to move the aircraft away from the gate.

Pushbacks at busy aerodromes are usually subject to ground control clearance to facilitate ground movement on taxiways. Once clearance is obtained, the pilot will communicate with the pushback tractor driver or a ground handler walking alongside the aircraft in some cases to start the pushback. To communicate, a headset may be connected near the nose gear.

Since the pilots cannot see what is behind the aircraft, steering is done by the pushback tractor driver and not by the pilots. Depending on the aircraft type and airline procedure, a bypass pin（前轮转向销）may be temporarily installed into the nose gear to disconnect it from the aircraft's normal steering mechanism.

Once the pushback is completed, the tow bar (拖杆) is disconnected, and the bypass pin is removed. The ground handler will show the bypass pin to the pilots to make it absolutely clear that it has been removed. The pushback is then complete, and the aircraft can start taxiing forward under its own power. (260 words)

Question 1   Why is reverse thrust not used to move aircraft backwards on the ground?

Question 2   What ground equipment are often used during pushback?

Question 3   What is often done when a pushback is completed?

VI. In this section, there are two passages. After you have read the two passages, there are five questions followed by four choices marked A, B, C and D. Decide which one is the most appropriate answer.

Passage One

The Takeoff and Departure Briefing

The takeoff and departure briefing is not as common as the arrival briefing, yet it's no less critical. In fact, given the pilot flying/pilot not flying discipline so widely accepted today, the departure crew briefing goes far to establish or clarify what's expected of crew members in their respective roles. This is the opportune time for the captain to establish or promote the atmosphere of enlightened leadership within the confines of the workplace. Even if the captain is the pilot not flying on a particular leg, the responsibilities of command dictate that he or she include amendments or modifications to the pilot flying's crew briefing should it be warranted (保证).

A typical takeoff and departure crew briefing should include: weather and related precautions, any special modification to the standard takeoff and initial climb profile, initial routing via designated SID and/or vector fix, initial clearance altitude including the transition altitude, com and nav radio configuration to include the assigned transponder code, and contingencies (意外情况) for critical emergency situations.

Weather conditions for your takeoff and departure should be part of every briefing. Special procedural precautions such as the use of engine or wing anti-ice, modified takeoff computations, or restrictions on the applicability of using a reduced power takeoff are examples of weather-related modifications best reviewed before getting involved in the actual activity.

Include in your briefing any special noise abatement takeoff and climb profile to be flown, a high terrain/obstacle clearance maneuver to execute, or a possible windshear complication. When briefing the departure climb, either by reference to a published SID or radar vector clearance, be sure to emphasize your initial clearance altitude. The nav radio setup should be noted, crossing restrictions discussed, and any special switchology or pilot not flying duties clarified.

Many different emergencies are possible during any takeoff and climb out. Reviewing the basics of "who will do what" in the event of a reject or engine failure is commonly briefed. Attempting to address further emergencies could run your briefing too long to be effective.

Once cleared onto the runway for takeoff, the pilot flying should re-brief the initial heading, route to be flown and the initial clearance altitude. He should also confirm the takeoff flap settings and respective takeoff flap settings and takeoff speeds. Conducting this last chance mini-brief will focus the crew on the details at hand. (388 words)

1. Till now, departure briefing seems _____.

    A. to be less important than arrival briefing

    B. to help make clear the pilots' specific roles

    C. to be no less common than arrival briefing

    D. to have become accepted among flight crew

2. When working as the PNF, the captain should _____.

    A. gives any orders to other crew members

    B. takes on his responsibilities as a leader

    C. ensures the proper lighting conditions

    D. modifies the departure briefing at will

3. Which of the following is not mentioned as one of the items of possible *related precautions*?

    A. Typhoon.   B. Altitude.   C. Navigation.   D. Route.

4. It is suggested in the passage that procedures of possible special weather conditions _____.

    A. be executed when they are encountered

    B. be briefed when the airplane is in climb

    C. be talked about in the course of departure briefing

    D. be modified at the captain's discretion as required

5. The initial clearance altitude should be re-briefed _____.

    A. when the aircraft is clear of the runway

    B. when the aircraft is permitted to line up

    C. when takeoff flap position is reconfirmed

    D. when the takeoff briefing is completed

Passage Two

Night Preparation and Preflight

Night flying requires that pilots be aware of, and operate within their abilities and limitations. Although careful planning of any flight is essential, night flying demands more attention to the details of preflight preparation and planning.

Preparation for a night flight should include a thorough review of the available weather reports and forecasts with particular attention given to temperature/dewpoint spread. A narrow temperature/dewpoint spread may indicate the possibility of ground fog. Emphasis should also be placed on wind direction and speed, since its effect on the airplane cannot be as easily detected at night as during the day.

On night cross-country flights, appropriate aeronautical charts should be selected, including the appropriate adjacent charts. Course lines should be drawn in black to be more distinguishable.

Prominently lighted checkpoints along the prepared course should be noted. Rotating beacons at airports, lighted obstructions, lights of cities or towns, and lights from major highway traffic all provide excellent visual checkpoints. The use of radio navigation aids and communication facilities add significantly to the safety and efficiency of night flying.

All personal equipment should be checked prior to flight to ensure proper functioning. It is very disconcerting to find, at the time of need, that a flashlight, for example, does not work. All airplane lights should be turned ON momentarily and checked for operation. Position lights can be checked for loose connections by tapping the light fixture. If the lights blink while being tapped, further investigation to determine the cause should be made prior to flight.

The parking ramp should be examined prior to entering the airplane. During the day, it is quite easy to see wheel chocks (轮挡), and other obstructions, but at night it is more difficult. A check of the area can prevent taxiing mishaps. (294 words)

6. During preflight preparation at night, the pilots should be more focused on _____.

    A. their own abilities   B. careful planning

    C. night conditions   D. preparation details

7. Wind conditions should be stressed during night preparation because _____.

    A. their influences are not easy to detect

    B. they may influence temperature at night

    C. they tend to change more than in the day

    D. they are more difficult to detect than in the day

8. Attention should be paid to obviously lighted checkpoints because _____.

    A. they tend to operate normally

    B. they are located near the airport

    C. they provide visual references

    D. they are independent of navigation aids

9. Prior to each flight, exterior lighting system should _____.

    A. be investigated   B. be examined

    C. be checked   D. be tapped

10. To examine the ramp can help the aircraft _____.

    A. be far away from stepladders

    B. not be affected by wheel chocks

C. have more night vision

D. be clear of obstacles on the ramp

VII. Read the following passage and retell it in your own words.

My Story about a Starter Problem

We did the Preflight Check, Before Start Check and **Before Taxi Checklist**[1] as per normal. During the taxi, the EGT indicator flashes. **EGT**[2] indicator is the engine temperature gauge. It's not supposed to flash during taxi. If it flashes during start, it means the engine might exceed its temperature limit. Flashing during taxi? That's not supposed to happen. I had this same situation last week in the same aircraft. Since we identify the problem as an indication problem, the captain told me to **reset**[3] the indicators. As I did that, everything went to normal. The whole flight proceeded as normal and reached destination without any problems.

After **disembarking**[4] the passengers, and preparing the plane for the next takeoff, we allowed the passengers of the next flight to board the plane which is going to KUL. We did everything as per normal but during the 2nd engine start (No.1 engine, left engine), there was no N2 indication. N2 is one of the fan inside the engine. It's supposed to spin and then turn the N1, the bigger fan outside. Since there was no indication, we **aborted**[5] the start. We opened the MEL and it told us to use the No.2 engine's timing and speed to estimate the indication for the No.1 engine that had lost its **tachometer**[6] display. (No.2 engine is on the right side of the aircraft if you're sitting inside it. We started No.2 engines first then the No.1 engine.)

So we shut down the No.2 Engine and restarted it to get its **timing**[7] and indication. At 26 seconds, fuel was introduced to the engine and at 1 min 13 seconds (13% N1), the starter should cut out. We were going to use these figures to start the other engine, the No.1 engine that had lost its tachometer display. This time when we monitored the N2, the indication was 0% N2 and the EGT didn't go down. Then the engineer shouted "Cut Cut" or "Stop Stop" or something like that. The captain rejected the start and then we knew something was wrong with the starter. Not only that, the Start Valve Open light was still **illuminated**[8]. It shouldn't. Because of this we couldn't turn on the

[1] 滑行前检查单

[2] 排气温度

[3] 复位

[4] 使……离机

[5] 中断

[6] 转速表

[7] 计时

[8] (灯)亮

Air-conditioning for the passengers.

To have a normal engine start, air from the APU is used to turn the starter, which is controlled by the Engine Start Switch. By placing this switch to **GRD**[9], it will open the Start Valve and air is used to spin the Starter Motor. Then the Starter Motor spins the N2. After 25% N2, fuel will be introduced so that it can run without the help of the starter. At 46% N2 indication, the starter should **disengage**[10] and the switch will automatically move to "OFF" position.

[9] 地面(位)

[10] 断开

The problem was that the Start Valve was stuck to Open position. The checklist (which we did earlier) want us to set the engine bleeds to ON and APU bleeds to OFF. That meant no conditioned air for the passengers. If we turned on the air-conditioning, the engine would start spinning because the start valve was open and air would spin the starter which would spin the N2 fan which might spin the N1 fan. (533 words)

Text B

Passage Rewriting

VIII. Read the following passage carefully and rewrite the passage under the given title with less than 200 words.

Executing Emergency Evacuation

_____

_____

Evacuation Slides

There are two types of aircraft evacuation slides: slides and rafts. A slide is for use only on land as a means of escape, although it has sufficient buoyancy to allow passengers to hold on to a lanyard running the length of it and use it as a buoyancy aid. A slide/raft is an evacuation slide that can be used both as a means of escape in a land evacuation and also as a life raft[1] in a landing on water. Slide/rafts usually feature an erectable canopy, outer compartments to hold passengers and survival packs containing items such as leak stoppers, paddles and flares.

Slides can also be single or dual lane, depending on the width of the exit, a dual lane slide being capable of evacuating a greater number of people quickly in an evacuation.

Slides and slide/rafts can be detached from the girt bar, usually by a two or three step procedures. This may, for example, involve lifting up the flap on the girt bar, and pulling the detach handle. These procedures are usually placarded red on the slide, "For Ditching Use Only". Once the slide is separated, the slide remains attached to the aircraft by a mooring line. This line will break if the airframe submerges, or can be disconnected with a pre-supplied knife or disconnect handle.

Before engine startup, all the aircraft doors are placed into the armed or automatic mode by the cabin crew. Methods of arming vary from aircraft to aircraft, but ultimately what is involved is the girt bar, a metal bar attached to the door end of the slide, being physically attached to brackets either in or adjacent to the door sill. On older aircraft, such as the Boeing 737, this is done physically by the cabin crew and on most other aircraft it involves pushing a lever on the door itself which arms the door internally.

If a rapid evacuation is required and the doors are opened while "armed", the opening of the door pulls the slide pack out of the bustle because the girt bar as physically attached to the aircraft floor. Due to the weight of both the door and the slide, great effort is involved in pushing the door open sufficiently to deploy the slide. Once the slide is completely free it will fall under gravity and after traveling a certain distance a pin will be pulled from a squib containing compressed gas and the slide will inflate. Should this system fail, the slide can be manually inflated by the cabin crew by pulling a manual inflation handle at the top of the slide. Should this also fail, standard operating procedures require the cabin crew to send passengers away from the door and to one that has a functioning escape slide.

Aircraft safety cards and in-flight safety demonstrations show the passengers where the nearest emergency exits are and how to use the evacuation slides. Additionally, Flight Attendants receive extensive safety training that covers the use of evacuation slides.

Inadvertent slide deployment occurs when the operator of the aircraft door attempts to open the door when it is in the armed position. A device can be used to prevent this problem. It works by sounding an audible alert (voice) when the door operator, whether trained or not, is about to open the door in the armed position. It works as an independent system, requiring no action other than arming the door as per normal standard operating procedures. When the door is placed in the armed position, the device is armed. It can be installed as a stand alone unit or integrated into the aircraft systems and powered from aircraft power. (611 words)

NEW WORDS

slide	/slaid/	n.	滑梯
raft	/rɑːft/	n.	安全筏
escape	/iˈskeip/	n.	逃离
buoyancy	/ˈbɔiənsi/	n.	浮力
lanyard	/ˈlænjəd/	n.	牵索
erectable	/iˈrektəbl/	adj.	可直立的
canopy	/ˈkænəpi/	n.	舱盖

paddle	/'pædl/	n.	滑桨，折动板
flare	/fleə/	v. & n.	拉平；电筒
detach	/di'tætʃ/	v.	分开；放下
dual	/'djuːəl/	adj.	双的，双重的
lane	/lein/	n.	道
placard	/'plækɑːd/	n.	标牌
ditch	/ditʃ/	v.	水上迫降
moor	/muə/	v.	抛锚
submerge	/səb'məːdʒ/	v.	浸没
armed	/ɑːmd/	adj.	预位的
bracket	/'brækit/	n.	支架
sill	/sil/	n.	梁；槛
bustle	/bʌsl/	n.	垫子
squib	/skwib/	n.	爆管
inflate	/in'fleit/	v.	充气
cabinet	/'kæbinit/	n.	橱柜
audible	/'ɔːdibəl/	a.	音频的；可听见的
revenue	/'revinjuː/	n.	(财政)收入

EXPRESSIONS

evacuation slide　撤离滑梯　　　　life raft　救生筏

girt bar　撑杆　　　　　　　　　emergency exit　紧急出口

NOTES

Life raft: A raft is an evacuation slide that can be both used as a means of land evacuation and also as a lift raft.

Lesson 11

Takeoff & Climb

Preview

One of the most fascinating parts of flying a plane is the takeoff. Takeoff is the phase of flight in which an aircraft goes through a transition from moving along the ground to flying in the air, usually starting on a runway. Text A gives us a brief description of the airplane takeoff and text B concerns the crosswind takeoffs and their differences from a normal takeoff.

Text A

Warming-up Activities

Picture Description

Please describe the following picture and be prepared to answer some questions.

Relevant Questions

1. Do you know anything about the airport regulations of bird management?
2. What dangers can birds bring to an aircraft taking off?
3. Please describe the normal takeoff procedure.
4. In your view, when are bird strikes more dangerous to aircraft? Why?

Takeoff

Takeoff is the phase of flight in which an aircraft goes through a transition from moving along the ground (taxiing) to flying in the air, usually starting on a runway. For balloons, helicopters and some specialized fixed-wing aircraft, no runway is needed.

For light aircraft, usually full power is used during takeoff. Large transport category (airliner) aircraft may use a reduced power for takeoff, where less than full power is applied in order to prolong engine life, reduce maintenance costs and reduce noise emissions. In some emergency cases, the power used can then be increased to increase the aircraft's performance. Before takeoff, the engines, particularly piston engines, are routinely run up at high power to check for engine-related problems. The aircraft is permitted to accelerate to rotation speed (Vr). The term rotation is used because the aircraft pivots around the axis of its main landing gear while still on the ground, usually due to manipulation of the flight controls to make this change in aircraft attitude.

The nose is raised to a nominal 5°-20° nose up pitch attitude to increase lift from the

wings and the airplane lifts off. For most aircraft, attempting a takeoff without a pitch-up would require cruise speeds while still on the runway.

Fixed-wing aircraft designed for high-speed operation have difficulty generating enough lift at the low speeds during takeoff. These are therefore fitted with high-lift devices, often including slats and usually flaps, which increase the camber of the wing, making it more effective at low speed, thus creating more lift. These are deployed from the wing prior to takeoff, and retracted during the climb. They can also be deployed at other times, such as prior to landing.

The speeds needed for takeoff are relative to the motion of the air (indicated airspeed). A headwind will reduce the ground speed needed for takeoff, as there is a greater flow of air over the wings. Typical takeoff air speeds for jetliners are in the 130-155 knot range (150-180 mph, 240-285 km/h). Light aircraft, such as a Cessna 150, take off at around 55 knots. Ultralights have even lower takeoff speeds. For a given aircraft, the takeoff speed is usually directly proportional to the aircraft weight; the heavier the weight, the greater the speed needed. Some aircraft specifically designed for short takeoff and landing can take off at speeds below 40 knots, and can even become airborne from a standing start when pointed into a sufficiently strong wind.

The takeoff speed required varies with air density, aircraft gross weight, and aircraft configuration (flap and/or slat position, as applicable). Air density is affected by factors such as field elevation and air temperature. This relationship between temperature, altitude, and air density can be expressed as a density altitude, or the altitude in the International Standard Atmosphere at which the air density would be equal to the actual air density.

Operations with transport category aircraft employ the concept of the takeoff V-Speeds, V_1, V_R and V_2. These speeds are determined not only by the above factors affecting takeoff performance, but also by the length and slope of the runway and any peculiar conditions, such as obstacles off the end of the runway. Below V_1, in case of critical failures, the takeoff should be aborted; above V_1 the pilot continues the takeoff and returns for landing. After the co-pilot calls V_1, he/she will call V_R or "rotate", marking speed at which to rotate the aircraft. The V_R for transport category aircraft is calculated such as to allow the aircraft to reach the regulatory screen height at V_2 with one engine failed. Then, V_2 (the safe takeoff speed) is called. This speed must be maintained after an engine failure to meet performance targets for rate of climb and angle of climb.

In a single-engine or light twin-engine aircraft, the pilot calculates the length of runway required to take off and clear any obstacles, to ensure sufficient runway to use for takeoff. A safety margin can be added to provide the option to stop on the runway in case of a rejected takeoff. In most such aircraft, any engine failure results in a rejected takeoff as a matter of course, since even overrunning the end of the runway is preferable to lifting off with insufficient power to maintain flight.

If an obstacle needs to be cleared, the pilot climbs at the speed for maximum climb angle (Vx), which results in the greatest altitude gain per unit of horizontal distance traveled. If no obstacle needs to be cleared, or after an obstacle is cleared, the pilot can accelerate to the best rate of climb speed (Vy), where the aircraft will gain the most altitude in the least amount of time. Generally speaking, Vx is a lower speed than Vy, and requires a higher pitch attitude to achieve. (806 words)

NEW WORDS

transition	/ˌtrænˈziʃən/	*n.*	过渡
prolong	/prəˈlɒŋ/	*v.*	延长
emission	/iˈmiʃən/	*n.*	释放
rotation	/rəʊˈteiʃən/	*n.*	转动；抬轮
pivot	/ˈpivət/	*n.*	枢轴
		v.	绕着枢轴转动
nominal	/ˈnɒminəl/	*adj.*	额定的
camber	/ˈkæmbə/	*n.*	曲度
deploy	/diˈplɔi/	*v.*	放出，张开
ultralight	/ˌʌltrəˈlait/	*n.*	超轻型飞机
proportion	/prəˈpɔːʃən/	*n.*	比例
density	/ˈdensiti/	*n.*	密度
slope	/sləʊp/	*n.*	斜坡；坡度
peculiar	/piˈkjuːljə/	*adj.*	特殊的
margin	/ˈmɑːdʒin/	*n.*	余度

EXPRESSIONS

air density　空气密度　　　　　　field elevation　机场标高

gross weight　总重　　　　　　　safety regulations　安全规章

screen height　屏高　　　　　　　run up　试车

NOTES

ISA　International Standard Atmosphere　国际标准大气

EXERCISES

Comprehension of Text A

I. Describe the given aviation terms from Text A in English.

1. takeoff
2. Rotation speed
3. V_1
4. V_2
5. rejected takeoff
6. takeoff distance

II. Answer the following questions after you have read Text A.

1. Do all takeoffs need runways? Give some examples.
2. Why is reduced-power takeoff used for some large airplanes?
3. How is engine runup performed before takeoff?
4. What pitch attitude is normally maintained at liftoff?
5. Why are high-lift devices installed on some fixed-wing aircraft?
6. Can you describe the effects wind conditions have on takeoffs of light aircraft?
7. What factors may affect takeoff speed?
8. What factors need to be considered during takeoff on the part of pilots?
9. What do aviation regulations state about takeoff distance and accelerate-stop distance?
10. What is the purpose of Landing and Takeoff Performance Monitoring Systems?

Reading Aloud

III. Read the following paragraph aloud until you can say it in a natural way from your memory.

Holding Pattern (等待航线)

A holding pattern for IFR aircraft is usually a racetrack pattern based on a holding fix. This fix can be a radio beacon such as an NDB or VOR. The fix is the start of the first turn of the racetrack pattern. Aircraft will fly towards the fix, and once there will enter a holding pattern. A standard one uses left-hand turns and takes about 4 minutes to complete. Deviations from this pattern can happen if long delays are expected; longer legs of two or three minutes may be used. Less frequent turns are more comfortable for passengers and crew. Additionally, right-hand turns may be assigned to some holding patterns if there are airspace restrictions nearby.

In the absence of a radio beacon, the holding fix can be any fixed point in the air, and can be created using two crossing VOR radials, an intersection or it can be at a specific distance from a VOR using a coupled DME. When DME is used, the inbound turn may be permanently defined by distance limits rather than in minutes. Furthermore, in appropriately equipped aircraft, GPS waypoints may be used to define the holding pattern, eliminating the need for ground-based navigational aids entirely. (201 words)

Vocabulary Practice

IV. Complete the following short passage by filling in the blanks with the words given in the box.

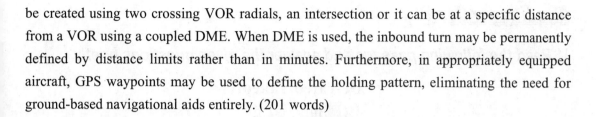

accident   rapidly   crashed   end   airframe   tail
result   repair   configuration   damage

Tailstrike (擦尾)

Tailstrike is an aviation term that describes an event in which the rear 1_____ of an aircraft touches (strikes) the runway. This can happen during takeoff of a fixed-wing aircraft if the pilot pulls up too 2_____ , leading to the rear end of the fuselage touching the runway. It can also occur during landing if the pilot flares too aggressively. This is often the 3_____of an attempt to land nearer to the runway threshold.

A tailstrike is physically possible only on an aircraft with tricycle landing gear; with a *tail dragger* 4_____, the tail is already on the ground. Some delta wing aircraft, which require a high angle of attack on takeoff, are fitted with small tailwheels to prevent tailstrikes. Tailstrike incidents rarely cause significant 5_____ or cause danger, but may cause financial losses as the planes have to be thoroughly inspected and repaired.

However, improper repair to the damaged 6_____after tailstrikes accidents may be responsible for fatal accidents that occur years later (including the worst single-aircraft 7_____as of 2008, the accident involving the *Japan Airlines Flight 123*) due to structural failure of the airframe at the site of the tailstrike after repeated cycles of pressurization and depressurization at the weak point of improper 8_____.

One tailstrike incident occurred on Sunday March 27, 1977. Seconds before KLM Flight 4805 9_____ into Pan Am Flight 1736, the captain of the KLM Boeing 747-200B, pulled the control column of the jet back too early, causing the horizontal stabilizer and 10_____ section to scrape the runway. He hoped to avoid a collision, but the drag and tailstrike caused the aircraft to rapidly lose speed. And when it was airborne, it was too late to avoid disaster. (287 words)

Reading for More

V. Read the following passage and answer the given questions briefly.

De-rated Takeoff
(减推力起飞)

Power setting values for maximum continuous thrust, go-around thrust, and full rated takeoff thrust are provided by AFM for both primary and alternate EEC modes. These power setting values are operational limits. Takeoff and go-around thrust ratings are limited to 5 minutes duration.

Power setting values for de-rated takeoff thrust are available from AFM for primary EEC mode. These power setting values are operational limits. Use of de-rated takeoff thrust is not allowed for alternate EEC mode operation. As a condition to the use of de-rated thrust, operators must perform periodic checks to ensure that the engines are capable of producing full takeoff thrust without exceeding any engine operating limits.

Operation at reduced takeoff thrust based on an assumed temperature higher than the actual ambient temperature is permissible if the airplane meets all applicable performance requirements at the planned takeoff weight and reduced thrust setting. The thrust reduction is referenced to the takeoff thrust, which in turn can be either full rated takeoff thrust or de-rated takeoff thrust. The amount of thrust reduction must not exceed 25 percent of the takeoff thrust. The power setting values for reduced takeoff thrust are available from AFM. At any time during the takeoff operation, the pilot may select takeoff thrust instead of reduced takeoff thrust. Use of reduced thrust procedures is allowed on a wet runway if suitable performance accountability is made for the increased stopping distance on the wet surface. Reduced thrust is not allowed when the runway is contaminated with standing water, ice, slush, or snow. Reduced thrust is not allowed for alternate EEC mode operation. When allowed, use of reduced thrust is at the pilot's discretion. (276 words)

Question 1 What is mentioned as one condition of the use of de-rated takeoff thrust?

Question 2 What is the limit to reduction of de-rated takeoff?

Question 3 How is de-rated takeoff made on contaminated runways?

VI. In this section, there are two passages. After you have read the two passages, there are five questions followed by four choices marked A, B, C and D. Decide which one is the most appropriate answer.

Passage One

Rejected Takeoff

A rejected takeoff (RTO) is a maneuver performed during the takeoff to expeditiously stop the airplane on the runway. As the airplane accelerates during takeoff roll, energy increases rapidly. The energy increase is in proportion to the square of the increase in speed. This energy must be dissipated to stop the airplane.

At low speeds, up to 80 knots, energy developed is not sufficient to cause difficulty in stopping the airplane.

As airspeed approaches V_1 for the balanced field condition, the effort required to stop the airplane on the runway for an RTO approaches maximum. After V_1, it may not be possible to stop the airplane on the runway. The decision to reject the takeoff must be made before V_1 so that the maneuver can be initiated no later than V_1 and must be accompanied by immediate accomplishment of the rejected takeoff maneuver.

Prior to V_1, a takeoff should be rejected in the event of engine failure, engine fire, unsafe configuration and any adverse condition significantly affecting the safety of flight. In the event any of the above occurs, the pilot not flying makes the appropriate announcement, such as "engine failure".

Simultaneously close the thrust lever (disengage the autothrottle, if required.) and apply maximum brakes. If RTO autobrakes are selected, monitor system performance and apply manual wheel brakes if the AUTOBRAKE DISARM light illuminates or deceleration is not adequate. Rapidly raise the speedbrakes and apply maximum reverse thrust consistent with conditions. Maintain reverse thrust and braking until runway length remaining permits transition to normal landing roll procedures.

Boeing recommends arming the RTO feature (if installed) on all takeoffs as it will ensure brake application early in the rejected takeoff.

Once thrust is set and takeoff roll has been established, rejecting a takeoff solely for illumination of the amber MASTER CAUTION light is not recommended. (302 words)

1. After he has decided to abort takeoff, the captain of an airplane shall _____.

    A. calculate the increase in airplane energy

    B. continue the acceleration of takeoff roll

    C. think about the existing runway surface conditions

    D. halt the airplane on the runway as rapidly as possible

2. What can be learned from the passage?

    A. Prior to 80 knots, there is much difficulty stopping the airplane.

B. After V1, it is not possible to stop the airplane on the runway.

C. Immediate rejected takeoff maneuver must begin at or before V1.

D. Prior to V_1, the effort required for an RTO exceeds maximum.

3. Should an engine failure occur before V1, the PNF shall _____.

    A. make call-outs   B. take over control

    C. take maneuvers   D. stop the airplane

4. During RTO, autobrakes should be switched to manual if _____.

    A. the autothrottle is not engaged   B. hydraulic pressure is available

    C. speed reduction is not sufficient   D. the reversers are not used in time

5. The duration of keeping using reverse thrust and brakes is determined by _____.

    A. the remaining runway length   B. the time of brake application

    C. normal landing roll procedures   D. MASTER CAUTION lights

Passage Two

Initial Climb

Once the proper pitch attitude is attained, it must be maintained. The initial climb after lift-off is done at this constant pitch attitude. Takeoff power is maintained and the airspeed allowed to accelerate. Landing gear retraction should be accomplished after a positive rate of climb has been established and confirmed. Remember that in some airplanes gear retraction may temporarily increase the airplane drag while landing gear doors open. Premature gear retraction may cause the airplane to settle back towards the runway surface. Remember also that because of ground effect, the vertical speed indicator and the altimeter may not show a positive climb until the airplane is 35 to 50 feet above the runway.

The climb pitch attitude should continue to be held and the airplane allowed to accelerate to flap retraction speed. However, the flaps should not be retracted until obstruction clearance altitude or 400 feet AGL has been passed. Ground effect and landing gear drag reduction results in rapid acceleration during this phase of the takeoff and climb. Airspeed, altitude, climb rate, attitude, and heading must be monitored carefully. When the airplane settles down to a steady climb, longitudinal stick forces can be trimmed out. If a turn must be made during this phase of flight, no more than 15° to 20° of bank should be used. Because of spiral (螺旋的) instability, and because at this point an accurate trim state on rudder and ailerons has not yet been achieved, the bank angle should be carefully monitored throughout the turn. If a power reduction must be made, pitch attitude should be reduced simultaneously and the airplane monitored carefully so as to preclude (防止) entry into an inadvertent descent. When the airplane has attained a steady climb at the appropriate en route climb speed, it can be trimmed about all axes and the autopilot engaged. (303 words)

6. During initial climb, _____.
   A. proper pitch attitude must be maintained
   B. pitch attitude may be kept constant
   C. gears are retracted at any rate of climb
   D. takeoff power can be reduced when airborne

7. Too early gear retraction may results in _____.
   A. change in pitch attitude   B. increase in airplane drag
   C. decrease in airplane lift   D. airplane backward movement

8. The altimeter begins to operate normally at 30-50 feet above the runway because of _____.
   A. induced drag   B. ground effect
   C. lag in instrument indication   D. improper configuration

9. Which of the following is NOT a condition to flap retraction?
   A. Pitch attitude remains constant.   B. Flap retraction speed is attained.
   C. Obstacle clearance is ensured.   D. 400 feet AGL at any airport.

10. If a turn is to be made during initial climb, the pilot shall _____.
   A. apply stick forces and make trims
   B. limit bank to 15° to 20° or above
   C. be aware of spiral instability in turn
   D. reduce power without pitch changes

VII. Read the following passage and retell it in your own words.

Engine Failure after Takeoff

The altitude available is, in many ways, the controlling factor in the successful accomplishment of an emergency landing. If an actual engine failure should occur immediately after takeoff and before a safe **maneuver**ing[1] altitude is attained, it is usually inadvisable to attempt to turn back to the field from where the takeoff was made. Instead, it is safer to immediately establish the proper **glide**[2] attitude, and select a field directly ahead or slightly to either side of the takeoff path.

[1] 机动

[2] 下滑

The decision to continue straight ahead is often difficult to make unless the problems involved in attempting to turn back are seriously considered. In the first place, the takeoff was in all probability made into the wind. To get back to the takeoff field, a downwind turn must be made. This increases the **groundspeed**[3] and rushes the pilot even more in the performance of procedures and in planning the landing approach. Secondly, the airplane will be losing considerable altitude during the turn and might still be in a bank when the ground is contacted. This would be a **catastrophe**[4] for the occupants, as well as the airplane. After turning downwind, the apparent increase in groundspeed could **mislead**[5] the pilot into attempting to slow down the airplane and cause it to stall. On the other hand, continuing straight ahead or making a slight turn allows the pilot more time to establish a safe landing attitude, and the landing can be made as slowly as possible, but more importantly, the airplane can be landed while under control.

[3] 地速

[4] 灾难

[5] 误导

To turn back to the runway following an engine failure on takeoff, the pilot should determine the minimum altitude an attempt of such a maneuver should be made in a particular airplane. Experimentation at a safe altitude should give the pilot an approximation of height lost in a descending 180° turn at idle power. By adding a safety factor of about 25 percent, the pilot should arrive at a practical decision height. The ability to make a 180° turn does not necessarily mean that the departure runway can be reached in a power-off glide; this depends on the wind, the distance traveled during the climb, the height reached, and the glide distance of the airplane without power. The pilot should also remember that a turn back to the departure runway may in fact require more than a 180° change in direction.

Consider the following example of an airplane which has taken off and climbed to an altitude of 300 feet AGL when the engine fails. After a typical 4 second reaction time, the pilot elects to turn back to the runway. Using a **standard rate** (3° change in direction per second) **turn**[6], it will take 1 minute to turn 180°. At a glide speed of 65 knots, the **radius**[7] of the turn is 2,100 feet, so at the completion of the turn, the airplane will be 4,200 feet to one side of the runway. The pilot must turn another 45° to head the airplane toward the runway. By this time the total change in direction is 225° equating to 75 seconds plus the 4 second reaction time. If the airplane in a poweroff glide descends at approximately 1,000 f.p.m., it will have descended 1,316 feet, placing it 1,016 feet below the runway. (547 words)

[6] 标准转弯

[7] 半径

Text B

Passage Rewriting

VIII. Read the following passage carefully and rewrite the passage under the given title with no less than 200 words.

How to Make a Safe Crosswind Takeoff

Crosswind Takeoff

Let's take a look at crosswind takeoffs and how they differ from a normal takeoff. We'll assume in this situation that we are taking off from runway 11 and the crosswind is from the northeast at about 10 mph. Thus, we have a crosswind blowing from left to right. The airplane used in this example will be a Piper J-3 Cub, a tailwheel airplane.

With our pre-takeoff checklist completed, we are ready to make our first crosswind takeoff.

First, taxi onto the runway and align the airplane with the centerline of the runway. Pull the control stick all the way back to the stop. This places our elevator in a full up position allowing for the relative wind to put a downward force on the tail. Remember, our only means of directional control at this point is with the steerable tail wheel.

Next, move the control stick all the way to the left into the direction from which the crosswind is coming. This will position our ailerons properly to compensate for the wind. The left aileron is in the up position while the right aileron is in the down position. We do this to help overcome the crosswind, as the left wing, the windward wing, will generate more lift than the leeward wing due to the fact that the leeward wing is partially blanked out by the fuselage. If we did not make this control input, the left wing would lift off before the right wing and we would find ourselves in a **precarious** position. At low speeds though, the ailerons will be ineffective until the plane picks up speed.

Now it's time to give some thought to the rudder and rudder pedals and how we'll use them during a crosswind takeoff. We can anticipate that the left-to-right crosswind will strike the vertical fin and rudder. In doing so, the wind will try to push the tail to the right forcing the nose to go left. Right rudder pressure will be needed. During the takeoff right rudder will also be needed to compensate for engine torque and propeller "P" factor, which try to push the nose to the left as well.

With a crosswind from the left, the left aileron is in the full up position and the right aileron is in the down position. Slight right rudder is applied.

After aligning the airplane with the centerline smoothly begin applying power. As the airplane begins picking up speed the aileron, elevator, and rudder controls begin becoming affective. With increasing ground speed right rudder will be needed for maintaining a straight track down the runway. Simultaneously ease the control stick slightly forward and remove about one-half of the left aileron correction. Remember, the faster we're moving, the more effective the controls, so it won't take as much aileron to keep the left wing from flying before the right wing.

As we ease the control stick forward, the tail wheel will lift from the runway. Under normal wind conditions we would want to keep the airplane in a tail low attitude, but in a crosswind we want to bring the tail up to a near level flight attitude.

Just as we reach liftoff speed, apply very slight backpressure on the control stick and move the ailerons to a neutral position. There, we've just lifted off the runway and we're now airborne. But our takeoff procedures are not yet quite done. After lifting off, relax the backpressure on the control stick and relax the rudder pressure you have been holding during the takeoff roll. This will allow the airplane to "seek" its own crab angle. The airplane's nose will actually move several degrees to the left. Once we've completed this step, ease back on the control stick and establish the normal 60 mph climb out speed. You've now executed a good crosswind takeoff!

Once comfortably airborne, relax the control pressures and allow the airplane to seek its own crab angle. Then continue a normal climb out.

The procedure for a crosswind takeoff in a tricycle gear aircraft is somewhat similar. After aligning the airplane with the runway centerline, turn the control yoke to the full left position, but leave the elevator in the neutral position. As you smoothly apply power, be prepared to apply right rudder pressure to maintain the straight track down the runway. While the groundspeed increases, about one-half of the aileron correction can be taken out. Hold the airplane on the runway until reaching liftoff speed. Then apply light back pressure on the yoke while returning the ailerons to neutral. Again, when airborne, let the airplane seek its own wind correction or crab angle. Follow this up with backpressure to establish the desired climb out speed. (785 words)

NEW WORDS

steerable	/ˈstiərəbl/	adj.	可操纵的
windward	/ˈwindwəd/	adj.	向风的；迎风的
leeward	/ˈliːwəd/	adj.	向下风的；背风的
precarious	/priˈkeəriəs/	adj.	危险的
torque	/tɔːk/	n.	扭力
crab	/kræb/	n.	侧航
yoke	/jəuk/	n.	操纵杆
smoothly	/smuːðli/	adv.	柔和地

EXPRESSIONS

blank out　取消，作废　　　　　　　pick up speed　加速

lift off　离地

Chart I – Departure

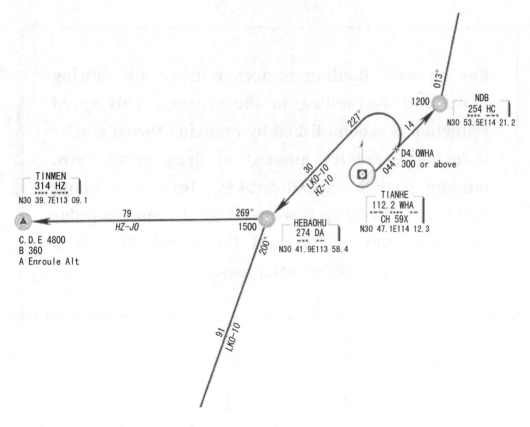

Lesson 12

Enroute & Approach

Preview

For aircraft, landing is accomplished by slowing down and descending to the runway. This speed reduction is accomplished by reducing thrust and/or inducing a greater amount of drag using flaps, landing gear or speed brakes. Text A is about Instrument Landing System (ILS), the most popular precision approach around the world. Text B is focused on IFR flight operations.

Text A

Warming-up Activities

Picture Description

Please describe the following picture and be prepared to answer some questions.

Relevant Questions

1. What are the common lighting systems in an airport?
2. What are the basic functions of the airport lighting systems?
3. Why is the Instrument Landing System (ILS) most common to airports around the world?
4. With routes stored in aircraft computers, do you agree that an aeronautical chart is still necessary in the cockpit? Why?

The Instrument Landing System

The Instrument Landing System (ILS) is a ground-based instrument approach system which provides precision guidance to an aircraft approaching a runway, using a combination of radio signals and, in many cases, high-intensity lighting arrays to enable a safe landing during Instrument Meteorological Conditions (IMC), such as low ceilings or reduced visibility due to fog, rain, or blowing snow.

Instrument Approach Procedure charts (or approach plates) are published for each ILS approach, providing pilots with the needed information to fly an ILS approach during Instrument flight rules (IFR) operations, including the radio frequencies used by the ILS components or navaids and the minimum visibility requirements prescribed for the specific approach.

An ILS consists of two independent sub-systems, one providing lateral guidance

(Localizer), the other vertical guidance (Glideslope or Glide Path) to aircraft approaching a runway. Aircraft guidance is provided by the ILS receivers in the aircraft by performing a modulation depth comparison.

A localizer (LOC) antenna array is normally located beyond the departure end of the runway and generally consists of several pairs of directional antennas. Two signals are transmitted on one out of 40 ILS channels between the frequency carrier frequency range 108.10 MHz and 111.95 MHz. One is modulated at 90 Hz, the other at 150 Hz and these are transmitted from separate but co-located antennas. Each antenna transmits a narrow beam, one slightly to the left of the runway centerline, the other to the right.

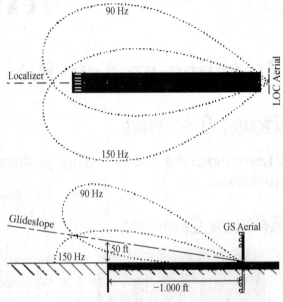

A glideslope or Glidepath (GP) antenna array is sited to one side of the runway touchdown zone. The GP signal is transmitted on a carrier frequency between 329.15 and 335 MHz using a technique similar to that of the localizer. The centerline of the glideslope signal is arranged to define a glideslope of approximately 3° above ground level.

Localizer and glideslope carrier frequencies are paired so that only one selection is required to tune both receivers.Their signals are displayed on an indicator in the instrument panel. This instrument is generally called the omni-bearing indicator or nav indicator. The pilot controls the aircraft so that the indications on the instrument remain centered on the display. This ensures the aircraft is following the ILS centreline. Vertical guidance, shown on the instrument by the glideslope indicator, aids the pilot in reaching the runway at the proper touchdown point. Some aircraft possess the ability to route signals into the autopilot, allowing the approach to be flown automatically by the autopilot.

In addition to the previously mentioned navigational signals, the localizer provides for ILS facility identification by periodically transmitting a 1020 Hz morse code identification signal. For example, the ILS for runway 04R at John F. Kennedy International Airport transmits IJFK to identify itself, while runway 04L is known as IHIQ. This lets users know the facility is operating normally and that they are tuned to the correct ILS. The glideslope transmits no identification signal, so ILS equipment relies on the localizer for identification.

Modern localizer antennas are highly directional. However, usage of older, less directional antennas allows a runway to have a non-precision approach called a localizer back course. This lets aircraft land use the signal transmitted from the back of the localizer array. This signal is reverse sensing so a pilot may have to fly opposite the needle indication. Highly directional antennas do not provide a sufficient signal to support a backcourse. In the United States, backcourse approaches are commonly associated with Category I systems at smaller airports that do not have an ILS on both ends of the primary runway.

At large airports, air traffic control will direct aircraft to the localizer via assigned headings, making sure aircraft do not get too close to each other, but also avoiding delay as much as possible. Several aircraft can be on the ILS at the same time, several miles apart. An aircraft that has intercepted both the localizer and the glideslope signal is said to be established on the approach. Typically, an aircraft will be established by 6 nautical miles (11 km) from the runway, or just after reaching the Final Approach Fix.

Once established on an approach, the Autoland system or pilot will follow the ILS and descend along the glideslope, until the Decision Altitude is reached (for a typical Category I ILS, this altitude is 200 feet above the runway). At this point, the pilot must have the runway or its approach lights in sight to continue the approach.

If neither can be seen, the approach must be aborted and a missed approach procedure will be performed. This is where the aircraft will climb back to a predetermined altitude and position. From there the pilot will either try the same approach again or divert to another airport. (782 word)

NEW WORDS

precision	/pri'siʒən/	*n.*	精密，精确
guidance	/'gaidəns/	*n.*	制导，引导
array	/ə'rei/	*n.*	系列；阵列
plate	/pleit/	*n.*	图
modulation	/ˌmɒdju'leiʃən/	*n.*	调制；调解；调节
reflection	/ri'flekʃən/	*n.*	折射；思考
aerial	/'eəriəl/	*n.*	天线
antenna	/æn'tenə/	*n.*	天线
predominance	/pri'dɒminəns/	*n.*	主宰，支配
tune	/tjuːn/	*v.*	调谐
omni-bearing	/ˌɒmni'beəriŋ/	*n.*	全方位
needle	/'niːdl/	*n.*	指针
abort	/ə'bɔːt/	*v.*	中断

| divert | /dai'vəːt/ | v. | 改航；备降 |

EXPRESSIONS

carrier frequency range　载频范围　　localizer back course　航向道反航道

morse code　莫尔斯编码　　coincide with　与……一致；巧合

NOTES

DDM	Difference in the Depth of Modulation	调制深度差
HSI	Horizontal Situation Indicator	水平状态指示器
CDI	Course deviation indicator	航道偏离指示器
FAF	Final Approach Fix	最后进近定位点

EXERCISES

Comprehension of Text A

I. Describe the given aviation terms from Text A in English.

1. precision guidance　　　　2. Instrument meteorological conditions

3. glideslope　　　　　　　　4. localizer receiver

5. vertical guidance　　　　　6. Final Approach Fix

II. Answer the following questions after you have read Text A.

1. What are the basic functions of the ILS?

2. What contents do Instrument Approach Procedure charts have?

3. What sub-systems does an ILS have?

4. What is the ground equipment of the instrument landing system?

5. What determines the difference between the two ILS localizer signals?

6. How are the ILS signals displayed in the airplane?

7. What is the basic operation principle of localizer back course?

8. How many categories does ILS have as far as you know?

9. What will a pilot do if the airplane is established on the localizer?

10. In what conditions must an ILS approach be aborted?

Reading Aloud

III. Read the following paragraph aloud until you can say it in a natural way from your memory.

Rapid Depressurization

Generally, a failure of the cabin pressurization system constitutes a pressurization emergency on board an aircraft. There are some main causes for such a failure: a malfunction of the pressure outflow, a malfunction of the pressure regulating valves, a physical leak in the system and damage to a door or window.

Some things happen in the cabin when rapid depressurization occurs. There are sudden temperature drop, fog and reduced visibility. Wind sucks items towards the hole and loose objects will fly about. Pressurization problems affect the human body in two significant ways: lack of oxygen to breathe and increased pressure within the body. Pressurization will affect communication with ATC because radio transmission will be poor owing to use of microphone within oxygen mask.

Immediate action should be taken by the pilot to combat pressurization problems by rapid immediate descent to FL 100-FL 150. At this level, There is enough air to breathe without oxygen masks. ATC should take some essential action in case of a pressurization emergency. Clear the airspace directly below the afflicted ACFT. ATC should also be ready to inform the pilot of minimum enroute altitude, if needed. (191 words)

Vocabulary Practice

V. Complete the following short passage by filling in the blanks with the words given in the box.

> stall   factor   elevator   relative   adjustment
> maneuvering   drag   bank   traffic   performance

Steep Turn (大坡度转弯)

Before starting the steep turn, the pilot should ensure that the area is clear of other air 1_____ since the rate of turn will be quite rapid. After establishing the manufacturer's recommended entry speed or the design 2_____speed, the airplane should be smoothly rolled into a selected 3_____angle between 45° to 60°. As the turn is being established, back-elevator pressure should be smoothly increased to increase the angle of attack. This

provides the additional wing lift required to compensate for the increasing load 4_____.

After the selected bank angle has been reached, the pilot will find that considerable force is required on the 5_____control to hold the airplane in level flight–to maintain altitude. Because of this increase in the force applied to the elevators, the load factor increases rapidly as the bank is increased. Additional back-elevator pressure increases the angle of attack, which results in an increase in 6_____. Consequently, power must be added to maintain the entry altitude and airspeed.

Eventually, as the bank approaches the airplane's maximum angle, the maximum 7_____or structural limit is being reached. If this limit is exceeded, the airplane will be subjected to excessive structural loads, and will lose altitude, or 8_____. The limit load factor must not be exceeded, to prevent structural damage.

During the turn, the pilot should not stare at any one object. To maintain altitude, as well as orientation, requires an awareness of the 9_____position of the nose, the horizon, the wings, and the amount of bank. The pilot who references the aircraft's turn by watching only the nose will have difficulty holding altitude constant; on the other hand, the pilot who watches the nose, the horizon, and the wings can usually hold altitude within a few feet. If the altitude begins to increase, or decrease, relaxing or increasing the back-elevator pressure will be required as appropriate. This may also require a power 10_____to maintain the selected airspeed. A small increase or decrease of 1° to 3° of bank angle may be used to control small altitude deviations. All bank angle changes should be done with coordinated use of aileron and rudder.

Short Answers

V. Read the following passage and answer the given questions briefly.

ILS Procedures with Radar

When radar is approved for approach control service, it is used to provide vectors to published instrument approach procedures, such as the ILS. Radar vectors can be provided for course guidance and for expediting traffic to the final approach course of any established instrument approach procedures.

Whenever ATC is providing radar vectors to the ILS approach course, you normally will be advised by the controller or through ATIS. The following is an example of a typical vector clearance. "*Mooney 782JM, descend and maintain 7,000, turn right heading 210, vector to the ILS runway 26R approach course.*"

As you get closer to the localizer, another vector and the approach clearance will be issued. For example, "*Mooney 782JM, position 11/2 miles east of the outer marker, turn right heading 257, maintain 7,000 until intercepting the localizer; cleared for ILS runway 26 R approach, contact tower 119.1 at Aruba."*

After receiving an approach clearance, you should maintain your last assigned altitude until you are established on a published segment of a route or instrument approach procedure. After you are established, published altitudes apply to descent within each succeeding segment, unless ATC assigns a different altitude. This is particularly important when you are receiving radar vectors or when you are operating on an unpublished route, since a premature descent could compromise your obstruction clearance. In addition, controllers are supposed to provide you with an altitude to maintain until you are established on a published segment of a route or instrument approach, or they may withhold the approach clearance until you are established.

If it becomes necessary for spacing, you may also be vectored across the approach course. If you determine that you are near the approach course and you have not been informed that you will be vectored across it, question the controller. You should not turn inbound on the final approach course unless you have received an approach clearance. Approach control will normally issue this clearance with the vector to intercept the final approach course, and the vector should enable you to establish your aircraft on the final approach course prior to reaching the final approach fix.

After passing the final approach fix inbound, you are expected to proceed direct to the airport and complete the approach, or to execute the published missed approach procedure. Radar service is automatically terminated when landing is completed. (391 words)

Question 1   What services can approach radars provide?

Question 2   What should a pilot do when an approach clearance is received?

Question 3   What is a pilot expected to do after passing final approach fix?

VI. Read the following passage and retell it in your own words.

Passage One

Decision Height

A decision height (DH) or decision altitude (DA) is a specified height or altitude in the precision approach at which a missed approach must be initiated if the required visual reference to continue the approach has not been acquired. This allows the pilot sufficient time to safely re-configure the aircraft to climb and execute the missed approach procedures while avoiding terrain and obstacles.

The decision height (DH) is the equivalent of the minimum descent altitude (MDA) for non-precision approaches; however, there are some significant differences. It is the level below which a pilot making such an approach must not allow his or her aircraft to descend unless the required visual reference to continue the approach has been established. Unlike an MDA, a missed approach need not be initiated once the aircraft has descended to the MDA, and that decision can be deferred to the missed approach point (MAP). So a pilot flying a non-precision approach may descend to the minimum descent altitude and maintain it until reaching the MAP, then initiate a missed approach if the required visual reference was not obtained. An aircraft must not descend below the MDA until visual reference is obtained, which differs from a DH in that an aircraft may descend below DH without visual reference so long as the missed approach procedure was initiated at or prior to the DH. For example, with a DH of 500ft MSL, it is legal for a pilot to allow the aircraft to descend to 450ft MSL if the missed approach procedure was initiated at or prior to 500ft. This would not be legal during a non-precision approach with a MDH of 500ft. This difference is due to the presence of vertical guidance during a precision approach, and thus terrain clearance near DH being less of an issue than near MDH during a non-precision approach.

If a runway has both precision and non-precision approaches defined, the MDA of the non-precision approach is almost always greater than the DA of the precision approach, due to the lack of vertical guidance of the non-precision approach: the actual difference will also depend on the accuracy of the navaid upon which the approach is based, with ADF approaches and SRAs tending to have the highest MDAs. (370 words)

1. Missed approach procedure should be executed when _____.

   A. visual reference requirements are not met

   B. terrain and obstacles are no longer a threat

   C. there is not sufficient time to make a landing

   D. the landing configuration is not established

2. DH differs from MDA in that MDA _____.

   A. is an altitude used in non-precision approach

   B. is the minimum altitude in all approaches

   C. is the minimum altitude an aircraft can descend to

   D. is reached with sufficient visual references

3. A pilot can fly to MAP at MDA _____.

   A. when landing configuration is established on final

   B. when missed approach procedure is initiated

   C. to see if there is sufficient visual reference to land

   D. until required visual reference is obtained at MAP

4. A pilot can descend below DH _____.

   A. if there is no visual reference at time

B. if missed approach procedure has begun

C. if precision vertical guidance is present

D. if the accuracy of the navaid is reliable

5. At runways with precision and non-precision approach, MDA is almost always greater than the DA because of _____.

A. vertical guidance of the non-precision approach

B. lack of vertical guidance of non-precision approach

C. reliability of the ground-based ADF equipment

D. almost sufficient terrain clearance at the airport

Passage Two

Final Approach

If operating in HDG SEL (航向选择方式), turn to position the drift angle pointer over the inbound course on the HIS. This will eliminate wind drift(偏流) effects during the final approach.

Prior to the FAF, set the altitude window to the nearest 100 feet increment above MDA, then select the V/S mode and an appropriate vertical speed. Less than 1,000 feet per minute may be selected if it is a long final approach. The objective is to arrive at MDA at a distance from the runway that will permit a normal 3 degree profile to the touchdown zone after maneuvering to the runway centerline.

With the altitude window set nearest one hundred above MDA, ALT ACQ (高度获得方式) engages high enough to permit a programmed rate of level off at the MCP altitude. This should be taken into account when selecting a vertical speed to descend to MDA, so that the MDA is reached prior to arrival at the visual descent point. If the weather conditions require a descent below the selected altitude to the published MDA, use CWS (驾驶盘操纵方式) or disconnect the autopilot and fly manually.

Reset the altitude window to the missed approach altitude after leveling off at or above MDA. The autopilot may be returned to CMD at the time if desired. Use of the autopilot is recommended if circling is required.

Maintain MDA until a 3 degree profile can be established to the touchdown zone. Approaching the point of visual profile intercept, establish the landing configuration. When leaving MDA, disconnect the autopilot and autothrottle and complete the landing. Both flight directors may be cycled OFF then ON after intercepting the visual profile to eliminate unneeded commands and permit normal TO/GA guidance should a go-around become necessary. (285 words)

6. During final approach, wind drift can be corrected _____.

    A. with autopilots operating in HDG SEL mode

    B. with the drift angle pointer properly positioned

    C. with the inbound course set during final approach

    D. with proper heading set with a heading selector

7. A suitable vertical speed should be selected _____.

    A. to reach MDA at proper profile and distance

    B. to establish a 3 degree profile prior to MDA

    C. to maintain 100 feet increment in V/S mode

    D. to ensure the longest distance to touchdown

8. What is said about level-off at MDA?

    A. The altitude window is set at one hundred feet.

    B. ALT ACQ engages at a suitably higher altitude.

    C. The MCP altitude is selected at or below MDA.

    D. The MDA is reached at the visual descent point.

9. CWS is most probably the abbreviation form of _____.

    A. Computer Warning System   B. Computer Wheel Speed

    C. Control Wheel System   D. Control Wheel Steering

10. Unnecessary commands should be removed with flight directors _____.

    A. to allow for normal landing guidance

    B. to prevent from possible changes in heading

    C. to provide guidance in case of going around

    D. to ensure the interception of descent profile

VII. Read the following passage and retell it in your own words.

IFR Flight Considerations

Approach practice is one of the most important aspects[1] of instrument [1] 方面
training. However, because it is normally conducted in VFR conditions, such
practice fails to provide some essential conditions for the instrument approach
under IFR. After you reach decision height or MDA, your instructor tells you
to take off the hood and land the airplane. You take off your hood, see the
runway in full view ahead, and land VFR as you have done many times before.

This type of training fails to expose you to the situations that you must
face in actual situations, such as seeing the ground but not the runway, or
faintly seeing approach lights oriented in a very strange way but no runway.

These conditions can lead to very serious problems and have led to many accidents. The final VFR segment of the instrument flight is the most hazardous, and many students are not exposed to these hazards during training.

The airlines recognize the dangers of the instrument approach to published **minimums**[2]. They do not permit pilots to execute an ILS approach to Category II minimums unless an autopilot or dual-display flight director is used. In addition, they normally require captains who are checking out in new airplanes to use "high minimums"– 100 feet above published minimums–for the first 100 hours in that airplane.

[2] 最低标准

You need to place similar personal minimums on your own instrument flying. If you fly once a week, you probably don't have the skill to do an approach to published minimums–and it is no good judgment to try to prove that you do. If you fly on instruments very rarely, you should raise your personal minimums farther. If you fly with equipment that is malfunctioning or of questionable reliability, you should raise your minimums even farther.

Preparing to fly an instrument approach is at least as important as actually flying it. This should include studying the approach and the missed approach procedures. Be sure to include the lighting aids for the runway you expect to use, so you will know what to expect when you break out of the clouds in low visibility. If you know the wind direction and speed, it gives you an idea of the angle and the position of the runway in the **windscreen**[3] – it will necessarily be straight ahead. Realize that as sparse visual cues become available, they can cause **disorientation**[4] until they are sufficiently clear to be used for landing.

[3] 风挡

[4] 迷航

Also know that there is a natural tendency for you to want to make a landing following an approach. This temptation can lead you to descend below minimums when even a few visual cues become available. However, when you do this, you are changing from the good quality cues that are presented on your flight instruments to uncertain outside visual cues that can cause confusion. When you first see the runway, use it only to confirm the instruments.

When decision height is reached, look outside and determine whether there is enough visibility to make a landing. If not, initiate a missed approach. If the approach can be continued, cross check the instruments and the outside environment to confirm agreement. If there is conflicting information between these two sources, execute the missed approach. (543 words)

Text B

Passage Writing

VIII. Read the following passage carefully and write a passage under the given title with no less than 200 words.

Flight in Low Visibility Conditions

_____

_____

Instrument Approaches

Instrument approaches are generally designed such that a pilot of an aircraft in instrument meteorological conditions (IMC), by the means of radio, GPS or INS navigation with no assistance from air traffic control, can navigate to the airport, hold in the vicinity of the airport if required, then fly to a position from where he or she can obtain sufficient visual reference of the runway for a safe landing to be made, or execute a missed approach if the visibility is below the minimums required to execute a safe landing. The whole of the approach is defined and published in this way so that aircraft can land if they suffer from radio failure; it also allows instrument approaches to be made procedurally at airports where air traffic control does not use radar or in the case of radar failure.

When aircraft are under radar control, air traffic controllers may replace some or all of these phases of the approach with radar vectors (the provision of headings on which the controller expects the pilot to navigate his aircraft) to the final approach, to allow traffic levels to be increased over those of which a fully procedural approach is capable. It is very common for air traffic controllers to vector aircraft to the final approach aid, e.g. the ILS, which is then used for the final approach. In the case of the rarely-used Ground-Controlled Approach (GCA), the instrumentation (normally Precision Approach Radar) is on the ground and monitored by a controller, who then relays precise instructions for adjustment of heading and altitude to the pilot in the approaching aircraft.

Low visibility approaches

Many instrument approaches allow for landing in conditions of low visibility. ICAO classifies ILS approaches as being in one of the following categories:

ILS Categories

Category	Decision Height (above threshold)	RVR limit
I	greater than 200 ft	550m or 1800 ft
II	less than 200 ft	350m or 1200 ft
IIIa	less than 100 ft	150m-200 m (see below)
IIIb	less than 50 ft	75m-150 m (see below)
IIIc	No DH	No RVR

Low visibility approaches are those in categories II and III.

For larger aircraft it is typical that these approaches are under the control of the flight control system with the flight crew providing a supervisory role.

Traditionally smaller aircraft which lacked redundancy in the flight control systems could not fly these approaches. A Head-Up Display allows the flight crew to fly the aircraft using the guidance cues from the ILS sensors so that if such a large deviation were seen, the pilot would be able to respond in an appropriate and safe manner. This is becoming increasingly popular with "feeder" airlines and most manufacturers of regional jets are now offering HUDs as either standard or optional equipment. In addition, a HUD can provide a low visibility take off capability.

For both automatic and HUD landing systems, the equipment requires special approval for its design and also for each individual installation. The design takes into consideration all of the additional safety requirements for operating an aircraft in close proximity to the ground. Once installed, the equipment also has additional maintenance requirements to ensure that it is fully capable of supporting reduced visibility operations. In all cases, additional crew training is required for such approaches, and a certain number of low visibility approaches must either be performed or simulated in a set period of time for pilots to stay "current" in performing them.

There are also air traffic control considerations with low visibility approaches: when using ILS, the integrity of the signal must be protected, which requires that certain areas of the airport close to the installations being free of other aircraft and vehicles. Also there must be bigger gaps between aircraft on final approach to both protect the ILS signal and to cope with slower runway vacation times. In addition, the airport itself has special considerations for low visibility operations including different lighting for approach, runways, and taxiways as well

as the location of emergency equipment. (670 words)

NEW WORDS

meteorological	/ˌmiːtiərəˈlɒdʒikəl/	adj.	气象的
align	/əˈlain/	v.	对准
vicinity	/viˈsiniti/	n.	附近
reference	/ˈrefərəns/	n.	参考；基准
relay	/riːˈlei/	n.	继电器
		v.	转递；传输
supervisory	/ˌsuːpəˈvaizəri/	adj.	监管的，管理的
feeder	/ˈfiːdə/	n.	支线
proximity	/prɒkˈsimiti/	n.	附近
jet	/dʒet/	n.	喷气机
gap	/gæp/	n.	间歇；间隔
intermediate	/ˌintəˈmiːdiət/	adj.	中间的
minima	/ˈminimə/	n.	最低值；最低标准
cue	/kjuː/	n.	提示；征候
vacation	/vəˈkeiʃən/	n.	脱离（跑道）

NOTES

| GPS | Global Positioning System | 全球定位系统 |
| INS | Inertial Navigation System | 惯性导航系统 |

Chart II a–Instrument Arrival

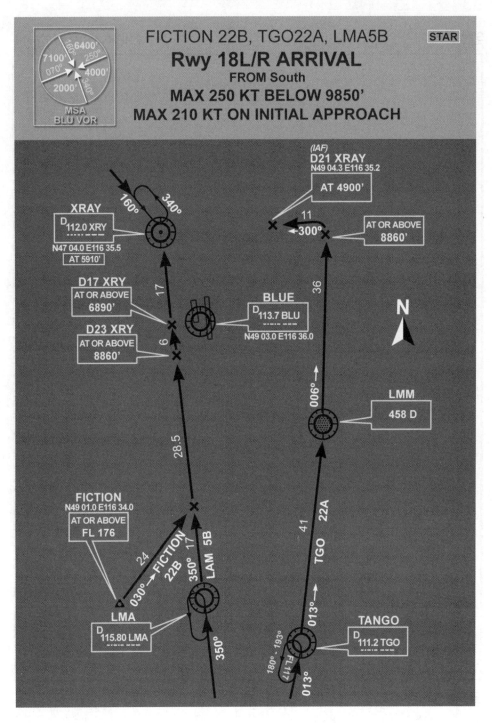

FICTION 22B, TGO22A, LMA5B — STAR

Rwy 18L/R ARRIVAL
FROM South
MAX 250 KT BELOW 9850'
MAX 210 KT ON INITIAL APPROACH

MSA BLU VOR

Chart II b–Instrument Arrival

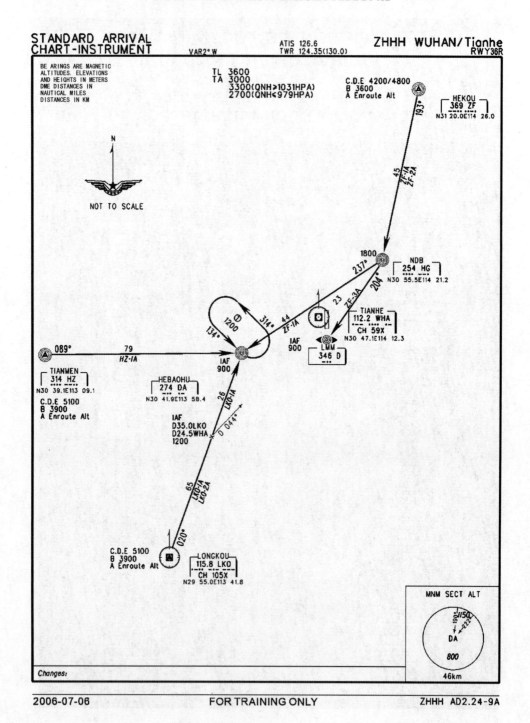

Lesson 13

Landing

Preview

> The landing phase of flight is crucial either in normal, abnormal or emergency situations, the last two of which impose threats on pilots as different conditions such as severe weather, mechanical problems, etc., may complicate the process of landing safely on the runway. Text A introduces us an emergency landing, belly landing, and text B is focused on crosswind landing.

Text A

Warming-up Activities

Picture Description

Please describe the following picture and be prepared to answer some questions.

Relevant Questions

1. Have you encountered any abnormal situation during landing?
2. What are the possible reasons for gear-up landing?
3. What factors should be taken into consideration prior to belly landing?
4. Are the standard procedures in normal landing still necessary in case of emergency landing? Give an example.

Gear-up Landing & Belly Landing

A belly landing or gear-up landing is when an aircraft lands without its landing gear fully extended and uses its underside, or belly, as its primary landing device. Normally the term gear-up landing refers to incidents in which the pilot forgets to extend the landing gear, while belly landing refers to incidents where a mechanical malfunction prevents the pilot from extending the landing gear.

During a belly landing, there is normally extensive damage to the airplane. Belly landings carry the risk that the aircraft may flip over, disintegrate, or catch fire if it lands too fast or too hard. Extreme precision is needed to ensure that the plane lands as straight and level as possible while maintaining enough airspeed to maintain control. Strong crosswinds, low visibility, damage to the airplane, or unresponsive instruments or controls greatly increase the danger of performing a belly landing. Still, belly landings are one of the most common types of aircraft accidents, and are normally not fatal if executed carefully.

The most common cause of gear-up landings is the pilot simply forgetting to extend the landing gear before touchdown. On any retractable gear aircraft, lowering the landing gear is part of the pilot's landing checklist, which also includes items such as setting the flaps, propeller and mixture controls for landing. Pilots who ritually perform such checklists before landing are less likely to land gear-up. However, some pilots neglect these checklists and perform the tasks by memory, increasing the chances of forgetting to lower the landing gear. Even experienced pilots are at risk, for they may be distracted and forget to perform the checklist or be interrupted in the middle of it by other duties such as collision avoidance or another emergency.

All aircraft with retractable landing gears are required to have a way to indicate the status of the landing gear, which is normally a set of lights that change colors from red to amber to green depending on whether the gear are up, in transit, or down. However, a distracted pilot may forget to look at these lights. This has led to aircraft designers building extra safety systems in the aircraft to reduce the possibility of human error. In small aircraft this most commonly takes the form of a warning light and horn which operate when any of the landing gear is not locked down and any of the engine throttles are retarded below a cruise power setting. However, the horn has been useless in situations when the pilot was unfamiliar with the aircraft and didn't know what the horn sounding was meant to indicate. Pilots have sometimes confused the landing gear warning horn with the stall warning horn. In other cases, pilots can't hear the horn on older aircraft due to wearing a modern noise-canceling headset.

In larger aircraft the warning system usually excludes the engine power setting and instead warns the pilot when the flaps are set for landing but the landing gear is not. An alternative system uses the ground proximity warning system or radar altimeter to engage a warning when the airplane is close to the ground and descending with the gear not down. Most airliners incorporate a voice message system which eliminates the ambiguity of a horn or buzzer and gives the pilot a clear verbal indication: "GEAR NOT DOWN". In addition, large aircraft are designed to be operated by two pilots working as a team. One flies the aircraft and handles communications and collision avoidance, while the other operates the aircraft systems. This provides a sort of human redundancy which reduces the workload placed on any one crew member, and provides for one crew member to be able to check the work of the other. The combination of advanced warning systems and effective crew training has made gear-up landing accidents in large aircraft extremely rare.

Mechanical failure is another cause of belly landings. Most landing gears are operated by electric motors or hydraulic actuators. Multiple redundancies are usually provided to prevent a single failure from failing the entire landing gear extension process. Whether electrically or hydraulically operated, the landing gear can usually be powered from multiple sources. In case

the power system fails, an emergency extension system is always available. This may take the form of a manually-operated crank or pump, or a mechanical free-fall mechanism which disengages the uplocks and allows the landing gear to fall due to gravity.

In cases where only one landing gear leg fails to extend, the pilot may choose to retract all the gear and perform a belly landing because he may believe it to be easier to control the aircraft during rollout with no gear at all than with one gear missing. (781 words)

NEW WORDS

belly	/ˈbeli/	n.	(机)腹
flip	/flip/	v.	翻转
disintegrate	/disˈintigreit/	v.	解体；瓦解
distract	/diˈstrækt/	v.	分心；分散注意力
amber	/ˈæmbə/	n.	琥珀色
transit	/ˈtrænsit/	v. & n.	过渡
retard	/riˈtɑːd/	v.	回收
horn	/hɔːn/	n.	喇叭
alternative	/ɔːlˈtəːnətiv/	adj.	备用的，备选的
ambiguity	/ˌæmbiˈgjuːiti/	n.	模糊
buzzer	/ˈbʌzə/	n.	蜂鸣器
redundancy	/riˈdʌndənsi/	n.	多于，余度
crank	/kræŋk/	n.	曲臂
uplock	/ˈʌplɔk/	n.	上位锁
rollout	/ˈrəulˈaut/	n.	滑跑
leg	/leg/	n.	臂；支柱

EXPRESSIONS

mixture control	混合比控制
ground proximity warning system	近地警告系统

EXERCISES

Comprehension of Text A

I. Describe the given aviation terms from Text A in English.

1. gear-up landing   2. belly landing

3. mixture control   4. landing checklist

II. Answer the following questions after you have read Text A.

1. What is the difference between gear-up landing and belly landing?

2. What damages does a belly landing do to an aircraft?

3. What is the primary reason for gear-up landing?

4. Why do some pilots forget to extend the landing gears?

5. What are the landing gear indications in the flight compartment?

6. In what situations can gear horn warning be useless?

7. What do you know about the gear warning system in larger aircraft?

8. What system or instrument can help landing gear warning?

9. What are the reasons for belly landing according to the passage?

10. Why do some pilots try to make a belly landing with only one gear leg problem?

Reading Aloud

III. Read the following paragraph aloud until you can say it in a natural way from your memory.

Hard Landing (硬着陆)

When the airplane contacts the ground during landings, its vertical speed is instantly reduced to zero. Unless provisions are made to slow this vertical speed and reduce the impact of touchdown, the force of contact may be so great that it could cause structural damage to the airplane. The purpose of pneumatic tires, shock absorbing landing gears, and other devices is to cushion the impact and to increase the time in which the airplane's vertical descent is stopped. Within a fraction of a second, the airplane must be slowed from this rate of vertical descent to zero, without damage.

During this time, the landing gear together with some aid from the lift of the wings must supply whatever force is needed to counteract the force of the airplane's inertia and weight. The lift decreases rapidly as the airplane's forward speed is decreased, and the force on the landing gear increases by the impact of touchdown. When the descent stops, the lift will be practically zero, leaving the landing gear alone to carry both the airplane's weight and inertia force. The load imposed at the instant of touchdown may easily be three or four times the actual weight of the airplane. (199 words)

Vocabulary Practice

IV. Complete the following short passage by filling the blanks with the words given in the box.

> location   technical   injury   assistance   hydraulics   sink
> operation   diversion   abnormal   flyable

Emergency Landing

An emergency landing is an unplanned landing made by an aircraft in response to a crisis which either interferes with the 1_____of the aircraft or involves sudden medical emergencies necessitating 2_____to the nearest airport.

There are several different types of emergency landings for powered aircraft: planned landing and unplanned landing.

For forced landing, the aircraft is forced to make a landing due to 3_____ problems, medical problems or weather conditions. Landing as soon as possible is a priority, no matter where. A forced landing may be necessary even if the aircraft is still 4_____. This can arise to either facilitate emergency medical or police 5_____or get the aircraft on the ground before a major system failure occurs which would force a crash landing or ditch situation.

Precautionary landing may result from a planned landing at a 6_____about which information is limited, from unanticipated changes during the flight, or from 7_____or even emergency situations. The sooner a pilot locates and inspects a potential landing site, the less the chance of additional limitations being imposed by worsening aircraft conditions, deteriorating weather, or other factors.

Crash landing is caused by the failure of or damage to vital systems such as engines, 8_____, or landing gear, and so a landing must be attempted where a runway is needed but none is available. The pilot is essentially trying to get the aircraft on the ground in a way which minimizes the possibility of 9_____ or death to the people aboard.

Ditching is the same as a crash landing, only on water. After the disabled aircraft makes contact with the surface of the water, the aircraft will typically 10_____ if it is not designed to float.

Reading for More

V. Read the following passage and answer the given questions briefly.

Factors Affecting Landing Distance

The field length requirements are contained in the performance section of the Operation Manual. Actual stopping distances for a maximum effort stop are approximately 60% of the field length requirement on a dry runway. Some factors that affect stopping distance are altitude and speed at the threshold, glide slope, use of thrust reversers, speedbrakes and brakes. Floating just off the runway surface before touchdown must be avoided, as this procedure uses a large portion of the available runway. The airplane should be landed as near the normal touchdown point as possible rather than be allowed to float in the air to bleed off speed. Airplane deceleration on the runway is about three times greater than in the air.

Height of the airplane over the end of the runway also has a very significant effect on total landing distance. For example, flying over the end of the runway at 100 feet altitude rather than 50 feet could increase the total landing distance by 950 feet on a 3 degree glide path. This change in total landing distance results primarily because of the length of runway used up before the airplane actually touches down. Glide path angle also affects total landing distance. Even while maintaining proper height over the end of the runway, total landing distance is increased as the approach path becomes flatter.

Stopping distance will also vary with wind conditions and any deviation form recommended approach speeds.

At high speeds, reverse thrust and speedbrakes drag are most effective. Make the speedbrake actuation and reverse thrust lever manipulation rapidly, with as little time delay as possible. (265 words)

Question 1   According to the passage, what are the factors affecting landing distance?

Question 2   Why should floating off the runway surface before touchdown be avoided?

Question 3   What is the author's attitude toward reverse thrust and speedbrakes at high speeds after touchdown? Why?

VI. In this section, there are two passages. After you have read the two passages, there are five questions followed by four choices marked A, B, C and D. Decide which one is the most appropriate answer.

Passage One

Touchdown

The touchdown is the gentle settling of the airplane onto the landing surface. The roundout and touchdown should be made with the engine idling and the airplane at minimum controllable airspeed, so that the airplane will touch down on the main gear at approximately stalling speed. As the airplane settles, the proper landing attitude is attained by application of whatever back-elevator pressure is necessary.

Some pilots may try to force or fly the airplane onto the ground without establishing the proper landing attitude. The airplane should never be flown on the runway with excessive speed. It is paradoxical (矛盾的) that the way to make an ideal landing is to try to hold the wheels a few inches off the ground as long as possible with the elevators. In most cases, when the wheels are within 2 or 3 feet off the ground, the airplane will still be settling too fast for a gentle touchdown; therefore, this descent must be retarded by further back-elevator pressure. Since the airplane is already close to its stalling speed and is settling, this added back-elevator pressure will only slow up the settling instead of stopping it. At the same time, it will result in the airplane touching the ground in the proper landing attitude and the main wheels touching down first, so that little or no weight is on the nosewheel.

After the main wheels make initial contact with the ground, back-elevator pressure should be held to maintain a positive angle of attack for aerodynamic braking, and to hold the nosewheel off the ground until the airplane decelerates. As the airplane's momentum (动能) decreases, back-elevator pressure may be gradually relaxed to allow the nosewheel to gently settle onto the runway. This will permit steering with the nosewheel. At the same time, it will cause a low angle of attack and negative lift on the wings to prevent floating or skipping, and will allow the full weight of the airplane to rest on the wheels for better braking action.

It is extremely important that the touchdown occuring with the airplane's longitudinal axis is exactly parallel to the direction in which the airplane is moving along the runway. Failure to accomplish this imposes severe side loads on the landing gear. To avoid these side stresses, the pilot should not allow the airplane to touch down while turned into the wind or drifting. (394 words)

1. During touchdown, the airplane is made at minimum controllable airspeed in order to _____.
    A. maintain engine at idle   B. approach stall speed
    C. control pitch attitude   D. settle down smoothly

2. The author seems to put an emphasis on _____.
    A. flying the airplane onto the ground
    B. forcing the airplane onto the ground
    C. proper landing speed at touch down

D. proper landing attitude at touchdown

3. The word "retard" (Line 9, Para. 2) most probably means _____.

    A. float   B. advance   C. slow   D. initiate

4. Proper pitch attitude is suggested during settling down with the intention of _____.

    A. relieving nose gear from weight   B. keeping stalling speed on landing

    C. the airplane touching the ground   D. reducing back-elevator pressure

5. The airplane's longitudinal axis should be parallel to the runway direction because _____.

    A. the airplane can counteract the crosswind effects

    B. the landing gear may be free from severe side loads

    C. landing direction at touchdown can be maintained

    D. floating or skipping at touchdown can be avoided

Passage Two

Flight Diversion

There will probably come a time when a pilot will not be able to make it to the planned destination. This can be the result of unpredicted weather conditions, a system malfunction, or poor preflight planning. In any case, the pilot will need to be able to safely and efficiently divert to an alternate destination. Before any cross-country flight, check the charts for airports or suitable landing areas along or near the route of flight. Also, check for navigational aids that can be used during a diversion.

Computing course, time, speed, and distance information in flight requires the same computations used during preflight planning. However, because of the limited cockpit space, and because attention must be divided between flying the airplane, making calculations, and scanning for other airplanes, take advantage of all possible shortcuts and rule-of-thumb (诀窍) computations.

When in flight, it is rarely practical to actually plot a course on a sectional chart and mark checkpoints and distances. Furthermore, because an alternate airport is usually not very far from your original course, actual plotting is seldom necessary.

A course to an alternate can be measured accurately with a protractor (量角器) or plotter (绘图器), but can also be measured with reasonable accuracy using a straight edge and the compass rose depicted around VOR stations. This approximation can be made on the basis of a radial from a nearby VOR or an airway that closely parallels the course to your alternate. However, remember that the magnetic heading associated with a VOR radial or printed airway is outbound from the station. To find the course to the station, it may be necessary to determine the reciprocal (反方向) of that heading. It is typically easier to navigate to an alternate airport that has a VOR or NDB facility on the field.

After selecting the most appropriate alternate, approximate the magnetic course to the alternate, using a compass rose or airway on the sectional chart. If time permits, try to start the diversion over a prominent ground feature. However, in an emergency, divert promptly toward your alternate. To complete all plotting, measuring, and computations involved before diverting to the alternate may only aggravate an actual emergency.

Once established on course, note the time, and then use the winds aloft nearest to your diversion point to calculate a heading and groundspeed. Once a groundspeed has been calculated, determine a new arrival time and fuel consumption. Give priority to flying the airplane while dividing attention between navigation and planning. When determining an altitude to use while diverting, consider cloud heights, winds, terrain, and radio reception. (424 words)

6. It is possible that a pilot _____.

    A. arrives at his destination airport at unscheduled time

    B. encounters some unexpected weather condition enroute

    C. needs to fly to his alternate airport for some reason

    D. checks the accuracy of navaids at the destination airport

7. In the second paragraph, the author argues that a pilot _____.

    A. makes calculations in flight as he does on the ground

    B. tries his best to find an effective way of calculation

    C. makes in-flight calculations as limited as possible

    D. pays his attention to scanning the flight instruments

8. Plotting a course on charts is hardly practical most probably because _____.

    A. aeronautical charts are no longer feasible in advanced aircraft

    B. protractor or plotter can hardly find its advantages without charts

    C. onboard VOR equipments have far more accuracy nowadays

    D. alternate airport is usually not very far from the original course

9. What is the author's attitude toward completing all plotting and computation in an emergency?

    A. Neutral.   B. Positive.   C. Negative.   D. Overstated.

10. When the airplane is established on course, the most important thing for the pilot to do is _____.

    A. to take notes of enroute time   B. to use wind conditions aloft

C. to calculate fuel consumption   D. to keep flying the airplane

VII. Read the following passage and retell it in your own words.

The Flare

The flare reduces the approach rate of descent to a more acceptable rate for touchdown. Unlike light airplanes, a jet airplane should be flown onto the runway rather than "held off" the surface as speed **dissipates**[1]. A jet airplane is aerodynamically clean even in the landing configuration, and its engines still produce residual thrust at **idle**[2] r.p.m. Holding it off during the flare in a attempt to make a smooth landing will greatly increase landing distance. A firm landing is normal and desirable. A firm landing does not mean a **hard landing**[3], but rather a deliberate or positive landing.

[1] 消散

[2] 慢车

[3] 硬着陆

For most airports, the airplane will pass over the end of the runway with the landing gear 30-45 feet above the surface, depending on the landing flap setting and the location of the touchdown zone. It will take 5-7 seconds from the time the airplane passes the end of the runway until touchdown. The flare is initiated by increasing the pitch attitude just enough to reduce the **sink rate**[4] to 100-200 feet per minute when the landing gear is approximately 15 feet above the runway surface. In most jet airplanes, this will require a pitch attitude increase of only 1° to 3°. The thrust is smoothly reduced to idle as the flare progresses.

[4] 下降率

The normal speed bleed off during the time between passing the end of the runway and touchdown is 5 knots. Most of the decrease occurs during the flare when thrust is reduced. If the flare is extended (held off) while an additional speed is bled off, hundreds or even thousands of feet of runway may be used up. The extended flare will also result in additional pitch attitude which may lead to a tail strike. It is, therefore, essential to fly the airplane onto the runway at the target touchdown point, even if the speed is excessive. A deliberate touchdown should be planned and practiced on every flight. A positive touchdown will help prevent an extended flare.

Pilots must learn the flare characteristics of each model of airplane they fly. The visual reference cues observed from each cockpit are different because window **geometry**[5] and visibility are different. The geometric relationship between the pilot's eye and the landing gear will be different for each make and model. It is essential that the flare maneuver be initiated at the proper

[5] 几何

height–not too high and not too low.

Beginning the flare too high or reducing the thrust too early may result in the airplane **floating**[6] beyond the target touchdown point or may include a rapid pitch up as the pilot attempts to prevent a high sink rate touchdown. This can lead to a tail strike. The flare that is initiated too late may result in a hard touchdown.

[6] 飘降；
平 飘

Proper thrust management through the flare is also important. In many jet airplanes, the engines produce a noticeable effect on pitch trim when the thrust setting is changed. A rapid change in the thrust setting requires a quick elevator response. If the thrust levers are moved to idle too quickly during the flare, the pilot must make rapid changes in pitch control. If the thrust levers are moved more slowly, the elevator input can be more easily coordinated. (539 words)

Text B

Passage Rewriting

VIII. Read the following passage carefully and rewrite the passage under the given title with no less than 200 words.

Safe Crosswind Landing

_____

_____

Crosswind Landing

A crosswind landing can best be summarized as follows: keep the windward wing down as long as possible to prevent sideward motion and perform the landing as any other. It sounds simple but executing is quite different.

There are two types of crosswind landings–the wing down and the crab method. The wing down method consists of keeping the airplane's longitudinal axis aligned with the flight path and the runway by lowering the windward wing while applying opposite rudder. If this sounds like the definition of a slip, it is precisely that. The crab method is done by keeping the airplane aligned with the centerline by establishing a crab angle. Then after crossing the runway threshold, rudder is applied to align the airplane with the centerline while simultaneously lowering the windward wing. Both crosswind approaches are acceptable, and both are practiced during crosswind flight training. It will be up to you to use whichever approach feels most comfortable to you.

Beginning with the wing down method, the procedure to follow is this. We'll assume we are landing on Runway 11 and the crosswind is from the northeast at 10 mph. After turning the airplane onto the final approach, lower the windward or left wing. Then apply right rudder to keep the airplane aligned with the centerline of the runway. Use only the amount of wing down needed to stop the drift of the airplane from left to right. Apply only enough right rudder to keep the airplane aligned with the centerline. Continue these control inputs right into the

flare and touch down. Depending on the crosswind velocity you may find yourself landing on the left main wheel and tail wheel. The right main wheel will remain off the ground for a few more seconds while the airplane continues to slow.

Without applying correct wing down and opposite rudder application, the airplane will drift to the side of the runway.

Once on the ground do not let up on the control stick backpressure. We need to keep the tail firmly planted on the runway for directional control. Continue moving the control stick all the way to the full left position as well, which will prevent the wind from picking up the left wing. Use whatever rudder inputs necessary to keep the airplane tracking down the centerline of the runway.

You may find during your approach to land, the crosswind will vary in velocity. This will require adjusting the left aileron and right rudder almost continuously during the approach. This may sound complicated but it really isn't once you've had an opportunity to try three or four crosswind landings.

When employing the crab technique, you will need to establish a crab angle on final approach to keep the airplane correctly aligned with the centerline of the runway. As you near the runway, prepare yourself to "kick" the airplane out of the crab and into the wing down attitude by applying right rudder and left aileron. From this point forward the remainder of the approach and touchdown is identical to the wing down method of crosswind landing.

Upon reaching the runway threshold, apply wing down and opposite rudder to keep the airplane aligned with the centerline.

Both crosswind landing methods are used comfortably in tricycle gear aircraft. The approaches are identical, but where they differ is at touchdown. Using the same example runway and wind situation, the tricycle airplane will first touch down on the left main wheel. Then ease the nose wheel down immediately after touchdown for effective ground steering in the crosswind. Seconds later the right main wheel touches down. Continue applying left aileron for the duration of the landing roll-out. (614 words)

NEW WORDS

sideward	/ˈsaidwəd/	adj.	侧向的
crab	/kræb/	n.	侧航
motion	/ˈməuʃən/	n.	运动
flare	/fleə/	n.	拉平
slip	/slip/	n.	侧滑
drift	/drift/	n.	偏流

Chart III–Instrument Landing

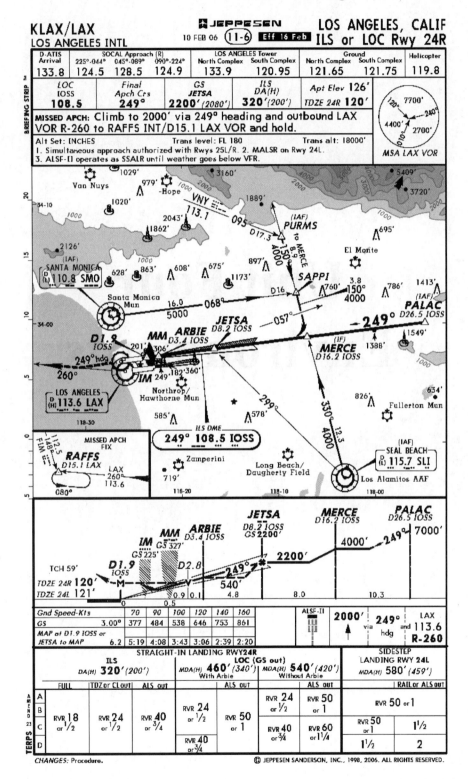

Unit Four
Environments

Lesson 14

Airport

Preview

An airport is a transportation center used for takeoffs and landings of aircraft to meet the needs of passengers and freight transportation. The selections in this lesson avail us of the major functions and general layout of airports, with the emphasis on the commercial airport and airport security. Other relevant knowledge is also covered, such as runway incursion avoidance, ground damage, etc.

Text A

Warming-up Activities

Picture Description

Please describe the following picture and be prepared to answer some questions.

Relevant Questions

1. Have you ever been to any airport?
2. How does a commercial airport differ from a general aviation airport?
3. Do you know anything about the general layout of an airport?
4. At which airport do you prefer to make a landing, a seaside airport or a highland airport? Tell me the reason for your option.

Airport

An airport is a transportation center used for the landing and takeoff of aircraft. Airports provide transportation not only for people but also for freight, such as mail, perishable foods, and other important items.

An airport is composed of several areas and structures that are designed to serve the needs of both aircraft and passengers. Runways are the long, narrow areas where airplanes take off and land. Taxiways are paths that aircraft follow from the runways to the terminal building, where passengers board and exit aircraft at areas called gates located within the terminal. The terminal also contains ticket and baggage counters. The control tower is located near the terminal. From this tower, people involved in air traffic control coordinate aircraft movement both in the air and on the ground. Maintenance and refueling facilities for aircraft are located near the runways or in nearby hangars. For security purposes, access to major

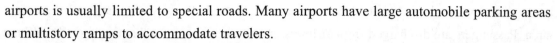

airports is usually limited to special roads. Many airports have large automobile parking areas or multistory ramps to accommodate travelers.

Airports differ in size and layout depending on their function and the types of aircraft that use them. There are three major types of airports: military airports, general aviation airports, and commercial airports. Military airports have one or two paved runways, generally 3,000 to 4,600 meters long. These airports are used only by military aircraft.

General aviation airports, which cater to small civilian aircraft, are smaller than commercial airports. They are often found in rural areas or in small towns. General aviation airports have one or two runways from 900 to 1,500 meters long. Some runways at general aviation airports are paved, but many are simply grass-covered paths. Facilities vary widely at general aviation airports, depending on the size of the airport.

Commercial airports are used by airlines. These airports may be small or large. Small commercial airports have one or two runways from 1,800 to 2,400 meters long and can accommodate larger aircraft than general aviation airports can. Large commercial airports serve the world's major cities. They usually have pairs of parallel runways from 3,000 to 3,700 meters in length. Airports approved as destinations for flights from other countries are known as international airports.

Commercial airports are designed to transfer passengers and freight to and from aircraft. In order to accomplish this transfer as efficiently and as safely as possible, airport operations are grouped into four general areas: aircraft services, passenger and freight services, support services, and airport security. Aircraft services focus on the flight, maintenance, and refueling of aircraft at the airport, as well as on air traffic control around the airport. Passenger services are centered in the terminal building, where passengers purchase tickets, load and retrieve baggage, and enter and exit aircraft. Terminals are designed in a variety of ways depending on the needs and size of a given airport. Airports also provide many support services indirectly related to air travel, such as restaurants, shops, parking, and aircraft emergency services.

Finally, airport security involves ensuring the safety of passengers and aircraft by screening passengers and their luggage for weapons or explosives. Airport security refers to the techniques and methods used in protecting airports and aircraft from crime. The goal of aviation security is to prevent harm to aircraft, passengers, and crew, as well as support national security and counter-terrorism policy.

Large numbers of people pass through airports. Such gatherings present a target for terrorism and other forms of crime due to the number of people located in a small area. Similarly, the high concentration of people on large airliners, the potential high lethality rate of attacks on aircraft, and the ability to use a hijacked airplane as a lethal weapon provide an alluring target for terrorism. Generally people are screened through airport security into the

concourses, where the exit gates to the aircraft are located. This area is often called a secure area. Passengers are discharged from airliners into the secure area so that they usually will not have to be re-screened if disembarking from a domestic flight; however they are still subject to search at any time. Eating establishments have started using plastic glasses and utensils as opposed to glasses made out of glass and utensils made out of metal to reduce the usefulness of such items as weapons.

Airport security provides defense by attempting to stop would-be attackers from bringing weapons or bombs into the airport. Airport security serves two purposes: to protect the airport from attacks and crime and to protect the aircraft from attacks. (752 words)

NEW WORDS

transportation	/ˌtrænspɔːˈteiʃən/	n.	交通；运输
freight	/freit/	n.	货物
perishable	/ˈperiʃəblə/	adj.	易腐烂的
terminal	/ˈtəːminəl/	n.	候机楼
exit	/ˈeksit/	n. & v.	出口；出去
gate	/geit/	n.	登机门
baggage	/ˈbægidʒ/	n.	行李
refuel	/riˈfjuəl/	v.	加油
hangar	/ˈhæŋə/	n.	机库
security	/siˈkjuəriti/		安全；保安
access	/ˈækses/		入口；进入
multistory	/ˌmʌltiˈstɔːri/	n.	多层
cater	/ˈkeitə/		配餐；一致
retrieve	/riˈtriːv/	v.	获得
explosive	/iksˈpləusiv/	n.	爆炸性物品
counter-terrorism	/ˌkauntəˈterərizm/		反恐怖
allure	/əˈljuə/	v.	诱惑
lethality	/liˈθæliti/	n.	致命
concourse	/ˈkɒŋkɔːs/	n.	大厅
disembark	/ˌdisimˈbɑːk/	v.	下飞机
utensil	/juːˈtensl/	n.	器皿

EXPRESSIONS

metric ton 公吨 secure area 安全区

EXERCISES

Comprehension of Text A

I. Describe the given aviation terms from Text A in English.

<ol>
<li>terminal building</li>
<li>commercial airport</li>
<li>secure area</li>
</ol>

2. gate
4. aviation security
6. counter-terrorism

II. Answer the following questions after you have read Text A.

1. What are the basic functions of an airport?

2. Can you describe the general areas and structures of an airport?

3. What do you know about commercial airports?

4. What areas does an airport have to serve the purpose of transferring passengers and cargos?

5. What are the support services that an airport provides?

6. What is the definition of airport security?

7. What are the goal and purposes of airport security?

8. Why is airport security so important?

9. What measures are taken to ensure airport security?

10. What does an airport do to defend against terrorism?

Reading Aloud

III. Read the following paragraph aloud until you can say it in a natural way from your memory.

John F. Kennedy International Airport

John F. Kennedy International Airport is an international airport located in the borough of Queens in New York City, about 12 miles (19 km) southeast of Lower Manhattan. In 2011 it was the busiest international air passenger gateway to the United States, handling more international traffic than any other airport in North America. It is also the leading freight gateway to the country by value of shipments. In 2011, the airport handled 47,809,910 passengers, making it the 17th busiest airport in the world and sixth busiest in the United States in terms of passenger traffic. The New York City metropolitan area's JFK International,

LaGuardia, and Newark International airports combine to create the largest airport system in the United States, second in the world in terms of passenger traffic, and first in the world in terms of total flight operations. In the last few years it has made extensive improvements to terminals, roadways and inter-terminal transportation. Over ninety airlines operate out of JFK. It is the base of operations for Jet Blue Airways and is a major international gateway hub for American Airlines and Delta Air Lines. In the past, it has been a hub for Eastern Air Lines, National Airlines, Pan American World Airways and Trans World Airlines. (214 words)

Vocabulary Practice

IV. Complete the following short passage by filling in the blanks with the words given in the box.

> tower   require   awareness   operations   incidents
>
> vehicle   collision   standard   areas   movement

Runway Incursion Avoidance
（避免跑道入侵）

A runway incursion is any occurrence at an airport involving an aircraft, 1_____, person, or object on the ground that creates a 2_____ hazard or results in a loss of separation with an aircraft taking off, landing, or intending to land. The three major 3_____ contributing to runway incursions are Communications, Airport knowledge, and Cockpit procedures for maintaining orientation.

Taxi operations 4_____ constant vigilance (警惕) by the entire flight crew, not just the pilot taxiing the airplane. This is especially true during flight training 5_____. Both the student pilot and the flight instructor need to be aware of the 6_____and operation of other aircraft and ground vehicles on the airport movement area. Many flight training activities are conducted at non-tower controlled airports. The absence of an operating airport control 7_____ creates a need for increased vigilance on the part of pilots operating at those airports.

Planning, clear communications, and enhanced situational awareness during airport surface operations will reduce the potential for surface 8_____. Safe aircraft operations can be accomplished and incidents eliminated if the pilot is trained early on and, throughout his/her flying career, accomplishes 9_____ taxi operating procedures and practices. This requires the development of the formalized teaching of safe operating practices during taxi operations. The flight instructor is the key to this teaching. The flight instructor should instill (灌输) in the student an 10_____ of the potential for runway incursion, and should emphasize the runway incursion avoidance procedures. (243 words)

Reading for More

V. Read the following passage and answer the given questions briefly.

Flight Dispatcher

A flight dispatcher is a person responsible for planning and monitoring the progress of an aircraft journey. In airline operations, depending on the type of certification the airline has, and depending on where the airline is based, both the pilot in command and the dispatcher are legally responsible for the safety of a flight. A dispatcher may have the authority to delay, divert or cancel a flight at any time, and a flight might not be able to be released without the signature of both the pilot in command and the dispatcher–again depending on the jurisdiction (管辖).

A dispatcher typically must be licensed by the aviation authority of a country. In order to obtain the license, the candidate must demonstrate extensive knowledge in meteorology and aviation, to a level that is comparable to that of an airline transport pilot license. The FAA ATPL written exam and the FAA Dispatcher written exam are identical; he/she is asked questions concerning safety of flight.

In the USA, 14 CFR PART 121 details the legal requirements governing dispatch release. After the release of a flight (in a joint responsibility environment) the dispatcher uses sophisticated software tools to monitor the flight's progress and advises the flight crew of any circumstances that might affect flight safety. Shared responsibility adds a layer of checks and balances to aircraft operation and greatly improves safety.

For airlines operating under 14 CFR PART 135, the dispatching duties and responsibilities are actually designated to flight followers. The main difference between a flight dispatcher and a flight follower* is that latter does not share legal responsibility for the operation of a flight. Also, followers are not required to attain a flight dispatcher's license, although they are usually encouraged to do so.

Many countries issue licenses which are based on ICAO Annex 1 and 6 as well as ICAO DOC 7192 D3, however not all countries have adopted a mandatory license and joint responsibility/flight watch operational control systems. The FAA has mandated the use of flight dispatchers/joint responsibility/flight watch since the "Civil Aeronautic Act" was passed in 1938. Canada has adopted a similar approach in the wake of a plane crash in Dryden,

Ontario in 1989.

Question 1　What are the duties and responsibilities of a flight dispatcher?

Question 2　How can a flight dispatcher get his license?

Question 3　What is the major difference between a flight dispatcher and a flight follower?

Note: A flight follower is responsible for performing administrative tasks maintaining awareness of weather and operational issues affecting flights and chartered aircraft, and assists in updating the flight/service deviations and coordination of shuttle operations.

VI. In this section, there are two passages. After you have read the two passages, there are five questions followed by four choices marked A, B, C and D. Decide which one is the most appropriate answer.

Passage One

Ground Damage

Aircraft are occasionally damaged by ground equipment at the airport. In the act of servicing the aircraft between flights, a great deal of ground equipment must operate in close proximity to the fuselage and wings. Occasionally the aircraft gets bumped or worse.

Damage may be in the form of simple scratches in the paint or small dents in the skin. However, because aircraft structures (including the outer skin) play such a critical role in the safe operation of a flight, all damage is inspected, measured and possibly tested to ensure that any damage is within safe tolerances. A dent that may look no worse than common "parking lot damage" to an automobile can be serious enough to ground an airplane until a repair can be made.

An example of the seriousness of this problem was the December 26, 2005 depressurization incident on Alaska Airlines flight 536. During ground services a baggage handler hit the side of the aircraft with a tug towing a train of baggage carts. This damaged the metal skin of the aircraft. This damage was not reported and the plane departed. Climbing through 26,000 feet (7,925 meters) the damaged section of the skin gave way due to the growing difference in pressure between the inside of the aircraft and the outside air. The cabin depressurized with a bang, frightening all aboard and necessitating a rapid descent back to denser (breathable) air and an emergency landing. Post landing examination of the fuselage revealed a 12 in × 6 in hole between the middle and forward cargo doors on the right side of the airplane.

The three pieces of ground equipment that most frequently damage aircraft are the passenger boarding bridge, catering trucks, and cargo belt loaders (传送带) . However, any other equipment found on an airport ramp can damage an aircraft through careless use, high winds, mechanical failure, and so on. (320 words)

1. Aircraft are sometimes damaged by ground equipment because of _____.

   A. engine shutdown   B. engine run-up   C. close proximity   D. long wingspan

2. Any ground damage to the aircraft should be carefully inspected and checked _____.

   A. to protect the skin of the airplane

   B. to ensure it is within safe tolerance

   C. to make clear the specific reasons

   D. to follow the company's regulation

3. The dent and parking lot damage are compared in the passage to indicate _____.

   A. all ground damages should be thoroughly repaired

   B. ground damage should be treated to different degrees

   C. damages like from automobile can be negligible

   D. the strength of the aircraft structure is very important

4. The cabin depressurization of Alaska Airlines flight 536 stems ultimately from _____.

   A. collision with the baggage handler

   B. collision with the luggage tug

   C. great difference in cabin pressure

   D. the rapid descent to a lower altitude

5. It can be inferred from the passage that _____.

   A. improper use of ground equipment is the reason for ground damages

   B. various examinations after damage are more important

   C. proper loading and catering are essential to avoid ground damages

   D. damages may come from ground equipments for various reasons

Passage Two

Reducing the Effect of Bird Strikes

There are three approaches to reduce the effect of bird strikes. The vehicles can be designed to be more bird resistant, the birds can be moved out of the way of the vehicle, or the vehicle can be moved out of the way of the birds.

Most large commercial jet engines include design features that ensure they can shut down after ingesting a bird weighing up to 1.8 kg (4lb). The engine does not have to survive the ingestion, but just be safely shut down. This is a "stand alone" requirement, i.e., the engine must pass the test, not the aircraft. Multiple strikes (due to hitting a bird flock) on twin engine jet aircraft are very serious events because they can disable multiple aircraft systems, requiring emergency action to land the aircraft, as in the January 15, 2009 forced ditching of US Airways Flight 1549.

Modern jet aircraft structures must be able to withstand one 4-pound-bird collision; the empennage must withstand one 8-pound-bird collision. Cockpit windows on jet aircraft must be able to withstand one 4-pound-bird collision without yielding or spalling (裂开).

To reduce bird strikes on takeoff and landing, airports engage in bird management and control. There is no single solution that works for all situations. Birds have been noted for their adaptability and control methods may not remain effective for long.

This includes changes to habitat around the airport to reduce its attractiveness to birds. Vegetation which produces seeds, grasses which are favored by geese, man-made food, a favorite of gulls, all should be removed from the airport area. Trees and tall structures which serve as roosts at night for flocking birds or perches for raptors should be removed or modified to discourage bird use.

Other approaches try to scare away the birds using frightening devices, for example sounds, lights, pyrotechnics (烟火), radio-controlled airplanes, decoy (诱饵) animals, lasers, dogs and firearms (枪炮) are also occasionally employed. (317 words)

6. To make the aircraft bird resistant, its structure _____.

    A. should be improved in its strength   B. should be bettered in design

    C. should be easy to avoid birds   D. should be good enough in stress

7. The example of forced ditching is cited in the passage to show _____.

    A. pilot should take special care when near birds

    B. aircraft should be manufactured to be bird resistant

    C. multiple bird strikes are very dangerous

    D. aircraft with two engines are more bird-resistant

8. What is the bird collision requirement for the tail section of the airplane?

    A. 4-pound.   B. 8-pound.   C. 5-pound.   D. Not mentioned.

9. There may not be absolutely effective measures of bird management because_____.

    A. most birds are very adaptable themselves

    B. bird management is still on its initial stage

    C. aircraft designs are far more than satisfactory

D. more solutions are better than any single way

10. Some measures are taken around the airport environment in order to _____.

    A. reduce the quantity of man-made food

    B. make the airport less bird-attractive

    C. raise more animals such as dogs

    D. frighten away the migrating birds

VII. Read the following passage and retell it in your own words.

Airport and Navigation Lighting Aids

The **lighting**[1] systems used for airports, runways, obstructions, and other [1] 灯光 visual aids at night are other important aspects of night flying.

Lighted airports located away from **congested**[2] areas can be identified [2] 拥挤的 readily at night by the lights **outlining**[3] the runways. Airports located near or [3] 突出 within large cities are often difficult to identify in the **maze**[4] of lights. It is [4] 迷宫 important not only to know the exact location of an airport relative to the city, but also to be able to identify these airports by the characteristics of their lighting pattern.

Aeronautical lights are designed and installed in a variety of colors and configurations, each having its own purpose. Although some lights are used only during low ceiling and visibility conditions, this discussion includes only the lights that are fundamental to visual flight rules (VFR) night operation.

It is recommended that prior to a night flight, and particularly a **cross-country**[5] night flight, the pilot check the availability and status of [5] 转场 lighting systems at the destination airport. This information can be found on aeronautical **charts**[6] and in the Airport/Facility Directory. The status of each [6] 航图 facility can be determined by reviewing pertinent Notices to Airmen (NOTAMs).

A rotating beacon is used to indicate the location of most airports. The beacon rotates at a constant speed, thus producing what appears to be a series of light flashes at regular **intervals**[7]. These flashes may be one or two [7] 间隔 different colors that are used to identify various types of landing areas. Lighted civilian land airports alternate white and green.

Beacons producing red flashes indicate obstructions or areas considered hazardous to aerial navigation. Steady burning red lights are used to mark obstructions on or near airports and sometimes to **supplement**[8] flashing lights [8] 补充

on en route obstructions. High intensity flashing white lights are used to mark some supporting structures of overhead transmission lines that stretch across rivers, **chasm**s[9], and **gorges**[10]. These high intensity lights are also used to identify tall structures, such as chimneys and towers.

[9] 峡谷
[10] 峡谷

The basic runway lighting system consists of two straight parallel lines of runway-edge lights defining the lateral limits of the runway. These lights are aviation white, although aviation yellow may be substituted for a distance of 2,000 feet from the far end of the runway to indicate a caution zone. At some airports, the intensity of the runway-edge lights can be adjusted to satisfy the individual needs of the pilot. The length limits of the runway are defined by straight lines of lights across the runway ends. At some airports, the runway **threshold**[11] lights are aviation green, and the runway end lights are aviation red.

[11] 入口

At many airports, the taxiways are also lighted. A taxiway-edge lighting system consists of blue lights that outline the usable limits of taxi paths. (458 words)

Text B

Passage Rewriting

VIII. Read the following passage carefully and rewrite the passage under the given title with no less than 200 words.

An Impressive Runway

_____

_____

Runways

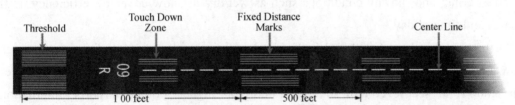

Threshold   Touch Down Zone   Fixed Distance Marks   Center Line

1 00 feet   500 feet

There are three types of runways: visual runways, non-precision instrument runways and precision instrument runways. Precision instrument runways, which are found at medium-and large-size airports, consist of a blast pad/stopway, threshold, designator, centerline, aiming point, and 500 ft, 1,000 ft/1,500 ft, 2,000 ft, 2,500 ft, and 3,000 ft touchdown zone marks. Precision runways provide both horizontal and vertical guidance for instrument approaches.

Runways may have different types on each end. To cut costs, many airports do not install precision guidance equipment on both ends. Runways with one precision end and any other type of end can install the full set of touchdown zones, even if some are past the midpoint. If a runway has precision markings on both ends, touchdown zones within 900 ft (274 m) of the midpoint are omitted, to avoid pilot confusion over which end the marking belongs to.

Runway pavement surface is prepared and maintained to maximize friction for wheel braking. To minimize hydroplaning following heavy rain, the pavement surface is usually grooved so that the surface water film flows into the grooves and the peaks between grooves

will still be in contact with the aircraft tires. To maintain the macro-texturing built into the runway by the grooves, maintenance crews engage in airfield rubber removal or hydro-cleaning in order to meet required FAA friction levels.

The active runway is the runway at an airport that is in use for takeoffs and landings. Since takeoffs and landings are usually done as close to "into the wind", headwind, as possible, wind direction generally determines the active runway.

Selection of the active runway, however, depends on a number of factors. At a non-towered airport, pilots usually select the runway most nearly aligned with the wind, but they are not obliged to use that particular runway. For example, a pilot arriving from the east may select to land straight into an east-west runway despite a minor tailwind or significant crosswind, in order to expedite his arrival, although it is recommended to always fly a regular traffic pattern to more safely merge with other aircraft.

At controlled airports, the active runway is usually determined by a tower supervisor. However, there may be constraints, such as policy from the airport manager (calm wind runway selection, for example, or noise abatement guidelines) that dictate an active runway selection that is not the one most nearly aligned with the wind.

At major airports, the active runway is based on weather conditions (visibility and ceiling, as well as wind, and runway conditions such as wet/dry or snow covered), efficiency, traffic demand, and time of day.

A runway of at least 6,000 ft (1,829 m) in length is usually adequate for aircraft weights below approximately 90,718 kg. Larger aircraft including widebodies will usually require at least 8,000 ft at sea level and somewhat more at higher altitude airports. International widebody flights, which carry substantial amounts of fuel and are therefore heavier, may also have landing requirements of 10,000 ft or more and takeoff requirements of 13,000 ft.

At sea level, 10,000 ft can be considered an adequate length to land virtually any aircraft. For example, at O'Hare International Airport, when landing simultaneously on 22R and 27L or parallel 27R, it is routine for arrivals from the Far East which would normally be vectored for 22R or 27R to request 27L.

An aircraft will need a longer runway at a higher altitude due to decreased density of air at higher altitudes, which reduces lift and engine power. An aircraft will also require a longer runway in hotter or more humid conditions. Most commercial aircraft carry manufacturer's tables showing the adjustments required for a given temperature. (610 words)

NEW WORDS

medium   /'miːdiəm/   *adj.*   中等的

blast	/'blɑːst/	n.	一股(强烈的气流)
pad	/pæd/	n.	平台；垫
stopway	/'stɒpwei/	n.	安全道
designator	/ˌdezig'neitə/	n.	符号；指示器
marking	/'mɑːkiŋ/	n.	标线
confusion	/kən'fjuːʒən/	n.	迷惑，困惑
pavement	/'peivmənt/	n.	道面
friction	/'frikʃən/	n.	摩擦
hydroplane	/'haidrəplein/	v.	滑水
groove	/gruːv/	n.	槽纹
merge	/məːdʒ/	v.	兼并，并入
macro-texture	/ˌmækrə'tekstʃə/	v.	高质地化
hydro-clean	/ˌhaidrə'kliːn/	v.	洁水

Chart IV – Airport Layout

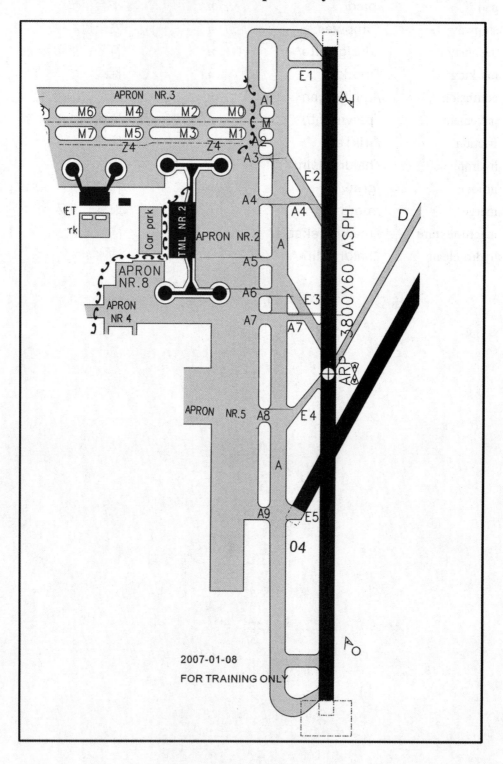

Lesson 15

Air Traffic Control

Preview

Air traffic control (ATC) is a service provided by ground-based controllers who direct aircraft on the ground and in the air. It plays an essential role in air traffic to ensure flight safety and facilitate the smooth flow of traffic. The selections in this lesson introduce us key concepts in ATC and the related areas to it as well as give us a brief picture of ATC in this world.

Text A

Warming-up Activities

Picture Description

Please describe the following picture and be prepared to answer some questions.

Relevant Questions

1. What does ATS refer to?
2. What are the common ATC faculties in China?
3. What services can pilots get from ATC?
4. How will you deal with the situation if a misunderstanding occurs in the communication between you and an ATC?

Area Control

ATC provides services to aircraft in flight between airports. Pilots fly under one of two sets of rules for separation: Visual Flight Rules (VFR) or Instrument Flight Rules (IFR). Air traffic controllers have different responsibilities to aircraft operating under the different sets of rules. While IFR flights are under positive control, in the US VFR pilots can request flight from air traffic control, which provides traffic advisory services on a time permitting basis and may also provide assistance in avoiding areas of weather and flight restrictions.

En-route air traffic controllers issue clearances and instructions for airborne aircraft, and pilots are required to comply with these instructions. En-route controllers also provide air traffic control services to many smaller airports around the country, including clearance off the ground and clearance for approach to an airport. Controllers adhere to a set of separation standards that define the minimum distance allowed between aircraft. These distances vary depending on the equipment and procedures used in providing ATC services.

En-route air traffic controllers work in facilities called Area Control Centers, each of which is commonly referred to as a "Center". Each center is responsible for many thousands of square miles of airspace (known as a Flight Information Region) and for the airports within that airspace. Centers control IFR aircraft from the time they depart from an airport or terminal area's airspace to the time they arrive at another airport or terminal area's airspace. Centers may also "pick up" VFR aircraft that are already airborne and integrate them into the IFR system. These aircraft must, however, remain VFR until the Center provides a clearance.

Center controllers are responsible for climbing the aircraft to their requested altitude while, at the same time, ensuring that the aircraft is  properly separated from all other aircraft in the immediate area. Additionally, the aircraft must be placed in a flow consistent with the aircraft's route of flight. This effort is complicated by crossing traffic, severe weather, special missions that require large airspace allocations, and traffic density. When the aircraft approaches its destination, the center is responsible for meeting altitude restrictions by specific points, as well as providing many destination airports with a traffic flow, which prohibits all of the arrivals being "bunched together". These "flow restrictions" often begin in the middle of the route, as controllers will position aircraft landing in the same destination so that when the aircraft are close to their destination they are sequenced.

As an aircraft reaches the boundary of a Center's control area, it is handed off or handed over to the next Area Control Center. In some cases this hand-off process involves a transfer of identification and details between controllers so that air traffic control services can be provided in a seamless manner; in other cases local agreements may allow silent handovers such that the receiving center does not require any co-ordination if traffic is presented in an agreed manner. After the hand-off, the aircraft is given a frequency change and begins talking to the next controller. This process continues until the aircraft is handed off to a terminal controller.

Since centers control a large airspace area, they will typically use long range radar that has the capability, at higher altitudes, to see aircraft within 200 nautical miles (370 km) of the radar antenna. They may also use TRACON radar data to control when it provides a better picture of the traffic or when it can fill in a portion of the area not covered by the long range radar.

In the US system, at higher altitudes, over 90% of the US airspace is covered by radar and often by multiple radar systems; however, coverage may be inconsistent at lower altitudes used by unpressurized aircraft due to high terrain or distance from radar facilities. A center may require numerous radar systems to cover the airspace assigned to them, and may also rely on pilot position reports from aircraft flying below the floor of radar coverage. This results in a large amount of data being available to the controller. To address this, automation systems have been designed that consolidate the radar data for the controller. This consolidation includes eliminating duplicate radar returns, ensuring the best radar for each geographical area is providing the data, and displaying the data in an effective format.

Centers also exercise control over traffic travelling over the world's ocean areas. These areas are also FIRs. Because there are no radar systems available for oceanic control, oceanic controllers provide ATC services using procedural control. These procedures use aircraft position reports, time, altitude, distance, and speed to ensure separation. Controllers record information on flight progress strips and in specially developed oceanic computer systems as aircraft report positions. This process requires that aircraft be separated by greater distances, which reduces the overall capacity for any given route. (801 words)

NEW WORDS

advisory	/əd'vaisəri/	*adj.*	咨询的
arrival	/ə'raivəl/	*n.*	进场；到达
restriction	/ri'strikʃən/	*n.*	限制
complicate	/'kɒmplikeit/	*v.*	使……复杂化
equivalent	/i'qwivələnt/	*adj.*	对等的；相等的
allocation	/ˌælə'keiʃən/	*n.*	分配
prohibit	/prə'hibit/	*v.*	禁止
square	/skweə/	*n.*	平方
terminal	/'tə:minəl/	*adj.*	终端的
		n.	终端；候机楼
flow	/fləu/	*n.*	流量
sequence	/'si:kwens/	*n.*	顺序
		v.	调序
seamless	/'si:mlis/	*adj.*	无缝的
co-ordination	/kəuˌɔ:di'neiʃən/	*n.*	协调
capability	/ˌkeipə'biliti/	*n.*	能力
coverage	/'kʌvəridʒ/	*n.*	覆盖范围
assign	/ə'sain/	*n.*	分配；指定
consolidate	/kən'sɔlideit/	*v.*	巩固；加强
eliminate	/i'limineit/	*v.*	消除
duplicate	/'dju:plikeit/	*adj.*	双重的

format	/'fɔːmæt/	*n.*	格式；形势
strip	/strip/	*n.*	条；带
capacity	/kə'pæsiti/	*n.*	容量

EXPRESSIONS

be consistent with	与……一致	hand off	移交
traffic flow	交通流量	radar return	雷达回波
long range	远程	FIR	飞行情报区
TRACON	终端雷达进近管制		

EXERCISES

Comprehension of Text A

I. Describe the given aviation terms from Text A in English.

1. ATC services
2. handover
3. flow restriction
4. procedure
5. radar coverage
6. area control

II. Answer the following questions after you have read Text A.

1. What are the differences between IFR flight and VFR flight according to the passage?

2. What do en-route controllers do according to this passage?

3. How can controllers define the minimum distance allowed between aircraft?

4. What is the area control center mainly responsible for?

5. Do you think the job of area controller is complicated or simple? Why?

6. When does handover take place in area control? And what gets involved in this process?

7. What is silent handover according to the passage?

8. What devices will be applied in area control?

9. Why do area controllers avail themselves of a large amount of data? And what can they do to deal with the data?

10. How is area control applied in ocean area?

Reading Aloud

III. Read the following paragraph aloud until you can say it in a natural way from your memory.

You may not deviate from a clearance unless you experience an emergency or the clearance will cause you to violate an FAR. When ATC issues a clearance, regulations specify that you are not to deviate from it, except in an emergency, unless an amended clearance is received or unless complying with that clearance will cause you to violate a rule. As a pilot in command, you must determine if you can safely comply with each clearance issued to you. Points to consider include whether compliance will cause you to violate a rule, such as being vectored into a cloud when you are operating VFR or to exceed the performance capabilities of yourself or your aircraft. If a clearance is unsafe or not appropriate, it is your responsibility to take whatever action is necessary to avoid a hazardous situation or a rule violation and promptly may compromise the safety of your aircraft. If you find it necessary to deviate from a clearance due to an emergency, you must notify ATC as soon as possible. When ATC has given you priority, as pilot in command, you may be requested to submit a written report within 48 hours to the chief of the ATC facility. (202 words)

Vocabulary

IV. Complete the following short passage by filling in the blanks with the words given in the box.

> congested   airports   ensure   weather   runway
> departures   issues   route   slot   releases

Clearance Delivery

Clearance Delivery is the position that 1_____ route clearances to aircraft, typically before they commence taxiing. These contain details of the 2_____ that the aircraft is expected to fly after departure. Clearance Delivery or, at busy 3_____, the Traffic Management Coordinator (TMC) will, if necessary, coordinate with the enroute center and national command center or flow control to obtain 4_____ for aircraft. Often, however, such releases are given automatically or are controlled by local agreements allowing "free-flow" 5_____. When weather or extremely high demand for a certain airport or airspace becomes a factor, Ground "slot delays" or re-routes (改航) may be necessary to 6_____ the system does not get overloaded. The primary responsibility of Clearance Delivery is to ensure that the aircraft have the proper route and 7_____ time. This information is also coordinated with the enroute center and Ground Control in order to ensure that the aircraft reaches the 8_____ in time to meet the slot time provided by the command center. At some airports, Clearance Delivery also plans aircraft pushbacks and engine starts, in which case it is known as the Ground Movement Planner (GMP): this position is particularly important at heavily 9_____ airports to prevent taxiway and apron gridlock(堵塞).

Flight Data (which is routinely combined with Clearance Delivery) is the position that is responsible for ensuring that both controllers and pilots have the most current information: pertinent 10_____ changes, outages(运行中断), airport ground delays/ground stops, runway closures, etc. Flight Data may inform the pilots using a recorded continuous loop on a specific frequency known as the Automatic Terminal Information Service (ATIS). (267 words)

Reading for More

V. Read the following passage and answer the given questions briefly.

Air traffic control (ATC) is a service provided by ground-based controllers who direct aircraft on the ground and in the air. The primary purpose of ATC systems worldwide is to separate aircraft to prevent collisions, to organize and expedite the flow of traffic, and to provide information and other support for pilots when able. In some countries, ATC may also play a security or defense role, or be run entirely by the military.

Preventing collisions is referred to as separation, which is a term used to prevent aircraft from coming too close to each other by use of lateral, vertical and longitudinal separation minima; many aircraft now have collision avoidance systems installed to act as a backup to ATC observation and instructions. In addition to its primary function, the ATC can provide additional services such as providing information to pilots, including weather and navigation information and NOTAMs (Notices to Airmen).

In many countries, ATC services are provided throughout the majority of airspace, and its services are available to all users (private, military, and commercial). When controllers are responsible for separating some or all aircraft, such airspace is called "controlled airspace" in contrast to "uncontrolled airspace" where aircraft may fly without the use of the air traffic control system. Depending on the type of flight and the class of airspace, ATC may issue instructions that pilots are required to follow, or merely flight information (in some countries known as advisories) to assist pilots operating in the airspace. In all cases, however, the pilot in command has final responsibility for the safety of the flight, and may deviate from ATC instructions in an emergency.

Although native language for the region is normally used, English language must be used if requested, as required by the International Civil Aviation Organization (ICAO.) (296 words)

Question 1 What are the main functions of ATC according to this passage?

Question 2 How can one distinguish controlled airspace from uncontrolled airspace?

Question 3 Who has the final responsibility for the safety of the flight?

VI. In this section, there are two passages. After you have read the two passages, there are five questions followed by four choices marked A, B, C and D. Decide which one is the most appropriate answer.

Passage One

Ground Control (sometimes known as Ground Movement Control abbreviated to GMC or Surface Movement Control abbreviated to SMC) is responsible for the airport "movement" areas, as well as areas not released (转租) to the airlines or other users. This generally includes all taxiways, inactive runways, holding areas, and some transitional aprons or intersections where aircraft arrive, having vacated the runway or departure gate. Exact areas and control responsibilities are clearly defined in local documents and agreements at each airport.

Any aircraft, vehicle, or person walking or working in these areas is required to have clearance from Ground Control. This is normally done via VHF/UHF radio, but there may be special cases where other processes are used. Most aircraft and airside vehicles have radios. Aircraft or vehicles without radios must respond to ATC instructions via aviation light signals or else be led by vehicles with radios. People working on the airport surface normally have a communication link through which they can communicate with Ground Control, commonly either by handheld radio or even cell phone. Ground Control is vital to the smooth operation of the airport, because this position impacts the sequencing of departure aircraft, affecting the safety and efficiency of the airport's operation.

Some busier airports have Surface Movement Radar (SMR) designed to display aircraft and vehicles on the ground. These are used by Ground Control as an additional tool to control ground traffic, particularly at night or in poor visibility. There are a wide range of capabilities on these systems as they are being modernized. Older systems will display a map of the airport and the target. Newer systems include the capability to display higher quality mapping, radar target, data blocks, and safety alerts, and to interface with other systems such as digital flight strips. (294 words)

1. Specific areas in the charge of Ground Controls _____.

    A. include airport "movement" areas   B. vary at different airports

C. can be leased to airlines   D. include active runway and aprons

2. It seems clear that Ground Control has the right to _____.

A. deny the entry of people or vehicle

B. modify the content of local documents

C. vacate an aircraft from active runways

D. equip some airport vehicles with radios

3. Workers on the airport surface must _____.

A. use VHF/UHF radios properly   B. work anywhere at the airport

C. communicate with Ground Control   D. ensure smooth operation of radios

4. Ground Control is of great importance in terms of the airport because it has influences on _____.

A. the proper use of handheld radios or cell phones

B. the normal operation of airborne radio equipments

C. the safe operation of people working at airports

D. the safety and efficiency of the airport's operation

5. Surface Movement Radar aims to _____.

A. control ground traffic, particularly at night

B. display aircraft and vehicles on the ground

C. enhance the capabilities on radar systems

D. display higher quality mapping or radar target

Passage Two

Many airports have a radar control facility that is associated with the airport. In most countries, this is referred to as Approach or Terminal Control; in the US, it is often still referred to as a TRACON (Terminal Radar Approach CONtrol) facility. While every airport varies, terminal controllers usually handle traffic in a 30 to 50 nautical mile (56 to 93 km) radius from the airport. Where there are many busy airports in close proximity, one single terminal control may service all the airports. The actual airspace boundaries and altitudes assigned to a terminal control are based on factors such as traffic flows, neighboring airports and terrain, and vary widely from airport to airport: a large and complex example is the London Terminal Control Centre which controls traffic for five main London airports up to 20,000 feet (6,100 m) and out to 100 nautical miles (190 km).

Terminal controllers are responsible for providing all ATC services within their airspace. Traffic flow is broadly divided into departures, arrivals, and overflights. As aircraft move in and out of the terminal airspace, they are handed off to the next appropriate control facility (a control tower, an en-route control facility, or a bordering terminal or approach control).

Terminal control is responsible for ensuring that aircraft are at an appropriate altitude when they are handed off, and that aircraft arrive at a suitable rate for landing.

Not all airports have a radar approach or terminal control available. In this case, the en-route center or a neighboring terminal or approach control may co-ordinate directly with the tower on the airport and vector inbound aircraft to a position from where they can land visually. At some of these airports, the tower may provide a non-radar procedural approach service to arriving aircraft handed over from a radar unit before they are visual to land. Some units also have a dedicated approach unit which can provide the procedural approach service either all the time or for any periods of radar outage for any reason. (335 words)

6. In spite of differences in airport radars, terminal controllers _____.

    A. service all the airports at close proximity

    B. change airspace boundaries and altitudes

    C. vary widely from airport to airport

    D. handle traffic at certain distance range

7. London Terminal Control is typical in that _____.

    A. it controls traffic for five main London airports

    B. it provides all ATC services within their airspace

    C. its has assigned airspace boundaries and altitudes

    D. traffic in this area is at higher flight levels

8. Terminal control has the responsibility of _____.

    A. providing ATC services within airport airspace

    B. sequencing traffic flow in bordering control

    C. making aircraft at proper altitudes on handover

    D. ensuring aircraft at a suitable altitude en route

9. Which of the following is indispensable to an airport without radar approach?

    A. En-route Control.   B. Terminal Control.

    C. Tower Control.   D. Area Control.

10. Which of the following is the best title of the passage?

    A. Radar Control.   B. Tower Control.

    C. Airport Controls.   D. Terminal Control.

VII. Read the following passage and retell it in your own words.

The Cooperative ATS Concept

One of the first tasks of the **program**[1] was to produce a high-level   [1] 项目

description of the best uses of data link for **ATS**[2] and develop the idea of enhanced integration of the airborne components into the ATS processes. This high-level description is known as the EUROCONTROL Cooperative Air Traffic Services concept. This concept also includes an operational approach for the gradual introduction of **data link**[3] into air traffic services.

    [2] 空中交通服务

    [3] 数据联路

Cooperation is not a new concept to ATS. ATS has always been cooperative. However, the form of cooperation is very much dictated by the tools available to controllers and pilots. As far as workload is concerned, today's ATS might be described as "rigid ATS". Tasks and responsibilities are clearly assigned to each ATS actor and there is very little scope for transferring the workload from one actor to another. The introduction of data link and the associated automation allows us to move towards "cooperative ATS", which focuses on redistribution of the workload among the main ATS actors. This workload redistribution takes three forms. In its simplest form, the workload is transferred from humans to machines. For many years, a great variety of tools have been investigated with a view to reducing controller workload. The development of data communication introduces a new technology with the capability to automate a substantial proportion of controller workload, e.g., communications for a tactical controller (which generally represent around 50% of the total workload). In general, data communications can improve all traditional controller and pilot tools.

Another aspect of cooperation is related to the redistribution of tasks within a controller team. In a classical controller team composed of two controllers, a planner and a tactical controller, all tasks involving communication with pilots are assigned to the tactical controller. Data link gives the planner the opportunity to do some of the communication tasks normally carried out by the tactical controller. This opportunity offers numerous ways of redistributing the workload within a controller team.

Cooperative sharing of tasks is not confined to air traffic controllers. **Delegating**[4] certain controller tasks to pilots is also feasible. If pilots are provided with the right information and tools, they can execute certain tasks in a much more efficient way than controllers. New ATC instructions to delegate tasks to pilots using new

    [4] 代表；委派

data-link-based cockpit devices, such as the Cockpit Display of Traffic
Information (CDTI), are being developed and **validated**[5] right now by
the AGC Program. The extent to which delegation to pilots can be taken
is still very much an open issue. In line with the EUROCONTROL
ATM 2000+ Strategy, the AGC Programme will also be investigating
the operational feasibility of advanced levels of delegation, often
referred to as "**autonomous**[6] operations", where pilots undertake a
substantial proportion of the tasks currently performed by controllers.

    Data link provides the ability to adapt to changing situations and
different environments. The redistribution of tasks can be carried out
dynamically and in accordance with predetermined parameters. The
controller can define the start and end points in time for any delegated
task, and specific **constraints**[7] for the delegation. The
EUROCONTROL COOPATS concept defines a **flexible**[8] framework
for dynamic workload redistribution among the main ATS actors. (527
words)

[5] 使生效

[6] 自主的

[7] 限制
[8] 灵活的

Text B

Passage Rewriting

VIII. Read the following passage carefully and rewrite the passage under the given title with no less than 200 words.

How Can Separation Be Maintained in VFR flight?

Separation and Clearance

The distance by which an aircraft avoids obstacles or other aircraft is termed separation. The most important concept of IFR flying is that separation is maintained regardless of weather conditions. In controlled airspace, Air Traffic Control (ATC) separates IFR aircraft from obstacles and other aircraft using a flight clearance based on route, time, distance, speed, and altitude. ATC monitors IFR flights on radar, or through aircraft position reports in areas where radar coverage is not available. Aircraft position reports are sent as voice radio transmissions. Aircraft position reports are not necessary if ATC can see an aircraft on radar. In the United States, a flight operating under IFR is required to provide position reports if ATC advises a pilot that the plane is not visible on radar, using the phrase "radar contact lost".

In the United States, IFR flights in controlled airspace require an ATC clearance for each part of the flight. A clearance always specifies a clearance limit, which is the farthest the aircraft can fly without a new clearance. In addition, a clearance typically provides a heading or route to follow, altitude, and communication parameters, such as frequencies and transponder codes. An aircraft operating VFR must also obtain a clearance to enter class B airspace, and is required to maintain an assigned heading or altitude restriction as long as it does not conflict with the safe operation of the aircraft.

In uncontrolled airspace, IFR aircraft do not require clearances, and they separate themselves from each other by using charted minimum altitudes to avoid terrain and obstacles,

standard cruising altitudes to avoid aircraft flying in different directions, and radio reports over mandatory locations.

In the United States and the Southern Domestic Airspace of Canada (SDA), airspace from 18,000 to 60,000 feet (5,586 to 18,288 meters) is designated as class A, requiring an IFR clearance for all aircraft. In other countries class A airspace begins higher or lower. For example, in France class A airspace begins at 19,500 feet (5,850 meters).

In the United States even when on a filed IFR flight plan, if conditions permit the pilot is responsible to maintain a watch for, and avoid other air traffic and obstructions. Separation may also be referred to as "protection".

The main purpose of IFR is the safe operation of aircraft in Instrument Meteorological Conditions (IMC). The weather is considered to be IMC when it does not meet the minimum requirements for Visual Meteorological Conditions. To operate safely in IMC, a pilot controls the aircraft relying on flight instruments, and ATC provides separation.

It is important not to confuse IFR with IMC. The vast majority of IFR flying is conducted in Visual Meteorological Conditions (VMC). Any time a flight is operating in VMC, the crew is responsible for seeing and avoiding other traffic; however, since the flight is conducted under Instrument Flight Rules, ATC still provides separation services.

During flight under IFR, there are no visibility requirements, and as such flying through clouds is permitted. There are still minimum conditions that must be present in order for the aircraft to take off and land; these will vary according to the type of navigation aids available, the location and height of terrain and obstructions in the vicinity of the airport, equipment on the aircraft, and according to qualifications of the crew. For example, landing at mountain airports such as Reno (KRNO) offer significantly different instrument approaches for aircraft landing on the same runway, but from opposite directions. Aircraft approaching from the north must make visual contact with the airport at a higher altitude than a flight approaching from the south, because of rapidly rising terrain south of the airport. This higher altitude allows a flight crew to start a climb earlier in case landing is not feasible.

Although large airliners and, increasingly, smaller aircraft now carry their own terrain awareness and warning system TAWS, these are primarily backup systems providing a last layer of defense if a sequence of errors or omissions causes a dangerous situation.

Under IFR, the primary means of navigation are either via radio beacons on the ground, such as VORs and NDBs, or GPS. In areas of radar coverage, ATC may also assign headings to IFR aircraft, also known as radar vectors. Radar vectors are the primary method for ATC to provide separation between aircraft for landing.

Modern Flight Management Systems have evolved sufficiently to allow a crew to plan a flight not only as to route and altitude, but to specific time of arrival at specific locations. This

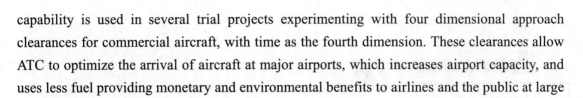

capability is used in several trial projects experimenting with four dimensional approach clearances for commercial aircraft, with time as the fourth dimension. These clearances allow ATC to optimize the arrival of aircraft at major airports, which increases airport capacity, and uses less fuel providing monetary and environmental benefits to airlines and the public at large respectively. (809 words)

NEW WORDS

separation	/ˌsepəˈreiʃən/	*n.*	间隔，间距
clearance	/ˈkliərəns/	*n.*	许可
concept	/ˈkɒnsept/	*n.*	概念
feasible	/ˈfiːzəbl/	*adj.*	可行的
mandatory	/ˈmændətəri/	*adj.*	强制的
evolve	/iˈvɒlv/	*v.*	发展；进化
optimize	/ˈɒptimaiz/	*v.*	使……最佳
monetary	/ˈmʌnəteri/	*adj.*	货币的；金融的
error	/ˈerə/	*n.*	错误；误差
omission	/əuˈmiʃən/	*n.*	省略
respective	/riˈspektiv/	*adj.*	各自的；分别的

Lesson 16

Weather

Preview

Various weather conditions are part of air travel, of which severe weather conditions like icing, wind shear, thunderstorms, typhoon etc. constitute a great threat to the flight safety of the aircraft operating in or near the area. The selections in this lesson familiarize us with some adverse weather conditions in relation to flight as well as IFR flying, by which separation is maintained regardless of meteorological visibility conditions.

Text A

Warming-up Activities

Picture Description

Please describe the following picture and be prepared to answer some questions.

Relevant Questions

1. What adverse weather conditions do you know about?
2. What threats can lightning pose to an aircraft in the air?
3. How can pilots get weather conditions prior to or during a flight?
4. Describe your experience of diversion because of severe weather condition.

Thunderstorms

Thunderstorms produce some of the most dangerous weather elements in aviation, and you should avoid penetrating them. There are three conditions necessary to create a thunderstorm – air that has a tendency toward instability, some type of lifting action, and relatively high moisture content.

The lifting action may be provided by several factors, such as rising terrain, fronts, or the heating of the earth's surface, convection. Thunderstorms progress three definite stages – cumulus, mature, and dissipating. You can anticipate the development of thunderstorms and the associated hazards by becoming familiar with the characteristics of each stage. Remember that other weather phenomena may prevent you from seeing their characteristics. For example, a cumulonimbus cloud may be embedded or obscured, by massive cloud layers.

In the cumulus stage, a lifting action initiates the vertical movement of air. As the air rises and cools to its dew-point, water vapor condenses into small water droplet or ice crystals.

If sufficient moisture is present, heat released by the condensing vapor provides energy for the continued vertical growth of the cloud. Because of strong updrafts, precipitation usually does not fall. Instead, the water drops or ice crystals rise and fall within the cloud, growing larger with each cycle. Updrafts as great as 3,000 f. p.m. may begin near the surface and extend well above the cloud top.

As the drops in the cloud grow too large to be supported by the updrafts, precipitation begins to fall to the surface. This creates a downward motion in the surrounding air and signals the beginning of the mature stage. The resulting downward draft may reach a velocity of 2,500 f.p.m. The down-rushing air spreads outward at the surface, producing a sharp drop in temperature, a rise in pressure, strong gusty winds, and wind front. As the thunderstorm advances, a rolling, turbulent, circular-shaped cloud may form at the lower leading edge of the cloud, roll cloud.

Mature Stage

As the mature stage progresses, more and more air aloft is disturbed by the falling drops. Eventually, the downdrafts begin to spread out within the cell, taking the place of the weakening updrafts. Because upward movement is necessary for condensation and the release of the latent energy, the entire thunderstorm begins to weaken. When the cell becomes an area of predominant downdrafts, it is considered to be in the dissipating stage. During this stage, the upper level winds often blow the top of the cloud downward, creating the familiar anvil shape. The anvil, however, does not necessary signal the storm's dissipation; severe weather can still occur well after its appearance.

Thunderstorms usually have similar physical features, but their intensity, degree of development, and associated weather do differ. They are generally classified as airmass or frontal storms. Airmass thunderstorms generally form in a warm, moist airmass and are associated with or scattered over a large area. These thunderstorms are most common during hot summer afternoons when winds are light or in coastal areas at night. Although they are usually scattered along individual mountain clouds, making them difficult to identify when approached from the windward side of a mountain. Nocturnal thunderstorms can occur in late spring and summer during the late night or early morning hours when westward, nocturnal storms cover many square miles, and their effects may continue for hours at a given location.

Thunderstorms typically contain many severe weather hazards, and may include lightning, hail, turbulence, gusty surface winds, or even tornadoes. These hazards are not confined to the cloud itself. For example, you can encounter turbulence in VFR conditions as far as 20 miles from the storm. It may help to think of a cumulonimbus cloud as the visible part of a widespread system of turbulence and other weather hazards. In fact, the cumulonimbus cloud is the most turbulent of all clouds. Indications of severe turbulence within the storm system include the cumulonimbus cloud itself, very frequent lightning, and roll clouds.

When conditions permit, you should circumnavigate thunderstorms. When the aircrew determines circumnavigation is possible, they merely alter course to take them around the storm. Since most individual thunderstorm cells are about five to ten miles in diameter, detouring to one side or another would not appreciably add to either the time or distance of the flight. In case of a line of thunderstorms, it is sometimes possible to circumnavigate them by flying through thin spots of precipitation between the storms. Exercise care in this procedure because another thunderstorm may lie on the other side of a thin spot.

When circumnavigation of thunderstorms is not possible, the next best course of action is to go over the top. Realize, thunderstorms build to great heights and this procedure is restricted to aircraft with the capability and fuel to climb to these altitudes. Some turbulence may be encountered in the clear air above the cloud. Additionally, hail can be thrown out the top of the cumulonimbus cloud. Thus, allow a margin of safety by choosing an altitude separation from the top of the thunderstorm of 1000 feet for every ten knots of wind speed at the altitude of the tops. Sometimes, aircraft cannot climb over the top of the cloud, but it will still be possible to fly over the saddlebacks between the build-ups. (867 words)

NEW WORDS

thunderstorm	/'θʌndəˌstɔːm/	n.	雷雨
penetrate	/'penitreit/	v.	穿越
instability	/ˌinstə'biliti/	n.	不稳定性
moisture	/'mɒistʃə/	n.	水分
front	/frʌnt/	n.	锋面
convection	/kən'vekʃən/	n.	对流
cumulus	/'kjuːmjuləs/	n.	积云
cumulonimbus	/ˌkjuːmjuləʊ'nimbəs/	n.	积雨云
embed	/im'bed/	v.	嵌入
dew-point	/'djuːpɔint/	n.	露点
condense	/kən'dens/	v.	凝结
droplet	/'drɒplit/	n.	水滴
crystal	/'kristəl/	n.	晶体
anvil	/'ænvil/	n.	砧状云
updraft	/'ʌpdrʌpt/	n.	上升气流

precipitation	/priˌsipiˈteiʃən/	*n.*	降水
gusty	/ˈgʌsti/	*adj.*	阵风的
aloft	/əˈlɒft/	*adj.*	高空的
cell	/sel/	*n.*	雷暴单体
airmass	/ˈeəmæs/	*n.*	气团
nocturnal	/nɒkˈtəːnl/	*adj.*	夜间的，晚上的
tornado	/tɔːˈneidəu/	*n.*	龙卷风
lightning	/ˈlaitniŋ/	*n.*	闪电
circumnavigate	/ˌsəːkəmˈnævigeit/	*v.*	绕航，绕飞
detour	/ˈdiːtuə/	*v.*	绕航，绕飞
saddleback	/ˈsædlbæk/	*n.*	鞍形背
build-up	/ˈbildˌʌp/	*n.*	雷雨

EXERCISES

Comprehension of Text A

I. Describe the given aviation terms from Text A in English.

1. the cumulus stage
2. precipitation
3. frontal thunderstorm
4. airmass storms
5. cumulonimbus cloud
6. circumnavigation

II. Answer the following questions after you have read Text A.

1. What conditions are necessary to create a thunderstorm?

2. What factors are needed to provide lifting forces for a thunderstorm?

3. What are the three stages of a thunderstorm?

4. Can you describe the development of precipitation?

5. How does a thunderstorm turn into a dissipating stage?

6. What characteristics do nocturnal thunderstorms have?

7. What types are frontal thunderstorms classified into?

8. What weather hazards go with a thunderstorm?

9. What should the flight crew do when there is a thunderstorm ahead of the route?

10. What should be kept in mind if the aircrew decide to overfly the thunderstorm?

Reading Aloud

III. Read the following paragraph aloud until you can say it in a natural way from your memory.

Ice in flight is bad news. It destroys the smooth flow of air, increasing drag while decreasing the ability of the airfoil to create lift. The actual weight of ice on an airplane is insignificant when compared to the airflow disruption it causes. As power is added to compensate for the additional drag and the nose is lifted to maintain altitude, the angle of attack is increased, allowing the underside of the wings and fuselage to accumulate additional ice. Ice accumulates on every exposed frontal surface of the airplane–not just on the wings, propeller, and windshield, but also on the antennas, vents, intakes, and cowlings. It builds in flight where no heat or boots can reach it. It can cause antennas to vibrate so severely that they break. In moderate to severe conditions, a light aircraft can become so iced up that continued flight is impossible. The airplane may stall at much higher speeds and lower angles of attack than normal. It can roll or pitch uncontrollably, and recovery might be impossible in some cases. Ice can also cause engine stoppage by either icing up the carburetor or, in the case of a fuel-injected engine, blocking the engine's air source. (201 words)

Vocabulary Practice

IV. Complete the following short passage by filling in the blanks with the words given in the box.

control	unsecured	inside	foreseen	aft
extreme	enter	streams	clear	moderate

Turbulence is air movement that normally cannot be seen. It may occur when the sky appears to be 1_____and can happen unexpectedly. It can be created by any number of different conditions, including atmospheric pressures, jet 2_____, mountain waves, cold or warm fronts, or thunderstorms. Turbulence can be divided into light turbulence, light chop, 3_____ turbulence, moderate chop, severe turbulence, 4_____turbulence and clear air turbulence.

In an extreme turbulence, aircraft is violently tossed about and is impossible to 5 _____. It may cause structural damage. The reactions 6_____ aircraft vary from occupants feeling slight strain against their seat belts and 7_____items being slightly displaced, through to occupants being forced violently against seat-belts, and unsecured items being tossed about.

(Imagine what it would be like if you were not wearing a seat belt!)

There are several notable problems with clear air turbulence. It cannot always be 8 _____ so there is no warning. It is usually felt at its mildest in the flight deck and is generally more severe in the 9_____ section. It can occur when no clouds are visible. Aircraft radars can't detect it. It is common at high altitudes, where cruising airline suddenly 10_____turbulent areas. Turbulence is the leading cause of in-flight injuries. There are countless reports of occupants who were seriously injured while moving about the passenger cabin when clear air turbulence is encountered. (231 words)

Reading for More

V. Read the following passage and answer the given questions briefly.

Protection from Lightning

Modern passenger jets have miles of wires and dozens of computers and other instruments that control everything from the engines to the passengers' music headsets. These computers, like all computers, are sometimes susceptible to upset from power surges. So, in addition to the design of the exterior of the aircraft, the lightning protection engineer must assure that no damaging surges or transients can be induced into the sensitive equipment inside of the aircraft. Lightning traveling on the exterior skin of an aircraft has the potential to induce transients into wires or equipment beneath the skin. These transients are called lightning indirect effects. Problems caused by indirect effects in cables and equipment are averted by careful shielding, grounding and the application of surge suppression devices when necessary. Every circuit and piece of equipment that is critical or essential to the safe flight and landing of an aircraft must be verified by the manufacturers to be protected against lightning in accordance with regulations of the FAA or a similar authority in the country of the aircraft's origin.

The other main area of concern is the fuel system, where even a tiny spark could be disastrous. Therefore, extreme precautions are taken to assure that lightning currents cannot cause sparks in any portion of an aircraft's fuel system. The aircraft skin around the fuel tanks must be thick enough to withstand a burn through. All the structural joints and fasteners must be tightly designed to prevent sparks as lightning current passes from one section to another. Access doors, fuel filler caps and any vents must be designed and tested to withstand lightning. All the pipes and fuel lines that carry fuel to the engines, and the engines themselves, must be verified to be protected against lightning. In addition, new fuels that produce less explosive vapors are now widely used.

Radomes are the nose cones of aircraft that contain radar and other flight instruments. The radome is an area of special concern for lightning protection engineers. In order to function, radar cannot be contained within a conductive enclosure. Protection is afforded by the application of lightning diverter strips along the outer surface of the radome. These strips are sized and spaced carefully according to simulated lightning attachment tests, while at the same time not significantly interfering with the radar. In many ways, diverter strips function like a lightning rod on a building. (397 words)

Question 1 What measures are taken by engineers to protect aircraft from lightning?

Question 2 Which systems of an aircraft are more subject to lightning?

Question 3 What is done to the radome as an area of special concern for lightning protection engineers?

VI. In this section, there are two passages. After you have read the two passages, there are five questions followed by four choices marked A, B, C and D. Decide which one is the most appropriate answer.

Passage One

Low Level Wind Shear

Wind shear is a sudden, drastic change in windspeed and/or direction over a very small area. Wind shear can subject an aircraft to violent updrafts and downdrafts as well as *abrupt* changes to the horizontal movement of the aircraft. While wind shear can occur at any altitude, low-level wind shear is especially hazardous due to the proximity of an aircraft to the ground. Directional wind changes of 180° and speed changes of 50 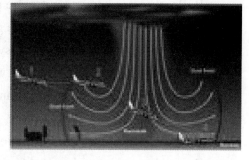 knots or more are associated with low-level wind shear. Low-level wind shear is commonly associated with passing frontal(峰面) systems, thunderstorms, and temperature inversions with strong upper level winds (greater than 25 knots).

Wind shear is dangerous to an aircraft for several reasons. The rapid change in wind direction and velocity changes the wind's relation to the aircraft, disrupting the normal flight attitude and performance of the aircraft. During a wind shear situation, the effects can be subtle or very dramatic depending on windspeed and direction of change. For example, a tailwind that quickly changes to a headwind will cause an increase in airspeed and performance. Conversely, when a headwind changes to a tailwind, the airspeed will rapidly decrease and there will be a corresponding decrease in performance. In either case, a pilot

must be prepared to react immediately to the changes to maintain control of the aircraft.

In general, the most severe type of low-level wind shear is associated with convective (对流的) precipitation or rain from thunderstorms. One critical type of shear associated with convective precipitation is known as a microburst. A typical microburst occurs in a space of less than 1 mile horizontally and within 1,000 feet vertically. The lifespan of a microburst is about 15 minutes during which it can produce downdrafts of up to 6,000 feet per minute. It can also produce a hazardous wind direction change of 45 knots or more, in a matter of seconds.

When encountered close to the ground, these excessive downdrafts (下沉气流) and rapid changes in wind direction can produce a situation in which it is difficult to control the aircraft. During an inadvertent takeoff into a microburst, the plane first experiences a performance-increasing headwind, followed by performance-decreasing downdrafts. Then the wind rapidly shears to a tailwind, and can result in terrain impact or flight dangerously close to the ground. (380 words)

1. "Abrupt" (Line 3, Para. 1) can be most appropriately replaced by _____.
   A. dangerous   B. rapid   C. dramatic   D. sudden
2. Why is a low level wind shear extremely dangerous?
   A. Because wind speed changes rapidly.
   B. Because wind direction is more unstable.
   C. Because airplane is close to the ground.
   D. Because it is related to frontal systems.
3. The hazard of a wind shear varies with _____.
   A. windspeed and direction of change
   B. flight attitude and performance
   C. strong upper level wind conditions
   D. increase in airspeed and performance
4. What is said about microburst?
   A. It often occurs at low levels in limited areas.
   B. It is associated with convective precipitation.
   C. Vertical change is more severe than horizontal.
   D. It takes place only in a matter of seconds.
5. Takeoff into a microburst is very dangerous because _____.
   A. more control maneuvers are required in this phase
   B. the behaviors of low level windshears are irregular
   C. the aircraft in takeoff are very closes to ground

D. ground effects get more severe with wind shears

Note: Microburst strong downdraft normally occurs over horizontal distances of 1 NM or less and vertical distances of less than 1,000 feet. In spite of its small horizontal scale, an intense microburst could induce wind speeds greater than 100 knots and downdrafts as strong as 6,000 feet per minute.

Passage Two

Clear Air Turbulence (CAT)

Clear air turbulence (晴空紊流) weather is the erratic (无常的) movement of air masses in the absence of any visual cues (征候), such as clouds. Clear air turbulence is caused when bodies of air moving at widely different speeds meet, at high altitudes (7,000-12,000 metres). This is frequently encountered around jet streams or sometimes near mountain ranges. Clear air turbulence is impossible to detect either with the naked eye or with conventional radar, meaning that it is difficult to avoid. However, it can be remotely detected with instruments that can measure turbulence with optical techniques, such as Doppler LIDARs.

This kind of turbulence creates a hazard for air navigation. Because aircraft move so quickly, they experience sudden unexpected accelerations or "bumps" as they rapidly cross invisible bodies of air which are moving vertically at many different speeds. Cabin crew and passengers on airliners have been injured (and in a small number of cases, killed, as in the case of a United Airlines Flight 826 on December 28, 1997) when tossed around inside an aircraft cabin during extreme turbulence.

Wake turbulence is another dangerous type of clear-air turbulence. The rotating vortex-pair created by the wings of a large aircraft can deflect or even flip a smaller aircraft on the ground. It can also lead to accidents in large aircraft as well. Delta Air Lines Flight 9570 crashed at the Greater Southwest International Airport in 1972 while landing behind a DC-10, leading to new rules for minimum following separation from "heavy" aircraft. American Airlines Flight 587 crashed shortly after takeoff from John F. Kennedy International Airport in 2001 due to pilot overreaction to wake turbulence from a Boeing 747 that caused separation of the vertical stabilizer. (295 words)

6. Clear air turbulence is frequently encountered _____.

    A. in moving air body   B. near mountain ranges

    C. in clear weather   D. near jet streams

7. Doppler LIDARs serves the purpose of _____.

    A. measuring wake turbulence   B. monitoring jet airplanes aloft

C. detecting turbulence optically   D. counteraction with turbulence

8. The example of United Airlines Flight 826 shows that _____.

    A. cabin crew and passengers should fasten their seat belts

    B. people in cabin should be prepared for possible CAT

    C. CAT constitutes an unexpected danger to air navigation

    D. invisible bodies of air are now controllable in airplanes

9. Wake turbulence differs from CAT in that it is dangerous _____.

    A. during cruise   B. during descent

    C. at low levels   D. at high levels

10. According to the passage, DC10 and B747 are similar because they _____.

    A. create clear air turbulence   B. generate wake turbulence

    C. are from large airports   D. were separated improperly

VII. Read the following passage and retell it in your own words.

Precipitation

Precipitation[1] refers to any form of water particles that form in the   [1] 降水
atmosphere and fall to the ground. It has a profound impact on flight safety.
Depending on the form of precipitation, it can reduce visibility, create icing
situations, and affect landing and takeoff performance of an aircraft.

Precipitation occurs because water or ice **particles**[2] in clouds grow in   [2] 颗粒
size until the atmosphere can no longer support them. It can occur in several
forms as it falls toward the Earth, including drizzle, rain, ice **pellets**[3], hail,   [3] 滴
and ice.

Drizzle is classified as very small water droplets, smaller than 0.02
inches in diameter. Drizzle usually accompanies fog or low stratus clouds.
Water droplets of larger size are referred to as rain. Rain that falls through
the atmosphere but **evaporates**[4] prior to striking the ground. Freezing rain   [4] 挥发
and freezing drizzle occur when the temperature of the surface is below
freezing; the rain freezes on contact with the cooler surface.

If rain falls through a temperature **inversion**[5], it may freeze as it passes   [5] 逆转
through the underlying cold air and fall to the ground in the form of ice
pellets. Ice pellets are an indication of a temperature inversion and that
freezing rain exists at a higher altitude. In the case of hail, freezing water
droplets are carried up and down by drafts inside clouds, growing larger in
size as they come in contact with more moisture. Once the updrafts can no

longer hold the freezing water, it falls to the Earth in the form of hail. Hail can be **pea**[6]-sized, or it can grow as large as 5 inches in diameter, larger than a softball. Snow is precipitation in the form of ice crystals that falls at a steady rate or in snow **showers**[7] that begin, change in intensity, and end rapidly. Falling snow also varies in size, being very small grains or large **flakes**[8].

[6] 豆

[7] 阵雪

[8] 片状

Snow grains are the equivalent of drizzle in size. Precipitation in any form poses a threat to safety of flight. Often, precipitation is accompanied by low **ceilings**[9] and reduced visibility. Aircraft that have ice, snow, or **frost**[10] on their surfaces must be carefully cleaned prior to beginning a flight because of the possible airflow disruption and loss of lift. Rain can contribute to water in the fuel tanks. Precipitation can create hazards on the runway surface itself, making takeoffs and landings difficult, if not impossible, due to snow, ice, or pooling water and very slick surfaces. (412 words)

[9] 云高

[10] 霜

Text B

Passage Rewriting

VIII. Read the following passage carefully and rewrite the passage under the given title with no less than 200 words.

Visual Reference in Flying

Instrument Meteorological Conditions

Instrument meteorological conditions (IMC) is an aviation flight category that describes weather conditions that require pilots to fly primarily by reference to instruments, and therefore under Instrument Flight Rules (IFR), rather than by outside visual references under Visual Flight Rules (VFR). Typically, this means flying in cloud or bad weather. Pilots sometimes train to fly in these conditions with the aid of products like Foggles, specialized glasses that restrict outside vision, forcing the student to rely on instrument indications only.

The most important concept of IFR flying is that separation is maintained regardless of meteorological visibility conditions. In controlled airspace, Air Traffic Control (ATC) separates IFR aircraft from obstacles and other IFR and known VFR aircraft by applying a flight clearance based on route, time, distance, speed, and altitude differences between aircraft. ATC monitors IFR flights by relying either on radar or aircraft position reports. Aircraft position reports are traditionally sent as voice radio transmissions, but increasingly also as electronic data exchanges. Aircraft position reports are not necessary if ATC has an aircraft in radar contact. In the United States a flight operating under IFR is required to fall back to position reports if advised radar contact lost.

IFR flights require an ATC clearance for each part of the flight. A clearance always specifies a clearance limit, which is the farthest the aircraft can fly without a new clearance. In addition, a clearance typically provides a heading or route to follow, altitude, and communication parameters, such as frequencies and transponder codes. An aircraft operating VFR must also obtain a clearance to enter class B and class C airspace, and is required to maintain an assigned heading or altitude restriction as long as it does not conflict with the safe operation of the aircraft.

The weather conditions required for flight under VFR are known as Visual Meteorological Conditions (VMC). IMC and VMC are mutually exclusive. In fact, instrument meteorological conditions are defined as less than the minima specified for visual meteorological conditions. The boundary criteria between VMC and IMC are known as the VMC minima, around which are known as "marginal VMC".

With good visibility, pilots can determine the attitude of the aircraft by utilizing visual cues from outside the aircraft, most significantly the horizon. Without such external visual cues, pilots must use an internal cue of attitude, which is provided by gyroscopically-driven instruments such as the Attitude Indicator (or "Artificial Horizon"). The availability of a good horizon cue is controlled by meteorological visibility, hence minimum visibility limits feature in the VMC minima. Visibility is also important in the avoidance of terrain.

Since the basic traffic avoidance principle of flying under Visual Flight Rules (VFR) is "see and avoid", it also follows that distance from cloud is an important factor in the VMC minima: as aircraft in cloud cannot be seen, a buffer zone from cloud is required.

ICAO recommends the VMC minima internationally; they are defined in national regulations, which rarely significantly vary from ICAO. The main variation is in the units of measurement as different states often use different units of measurement in aviation. The criteria tend to be stricter in controlled airspace, where there is a lot of traffic therefore greater visibility and cloud clearance is desirable. The degree of separation provided by Air Traffic Control is also a factor: for example in Class A and B airspace where all aircraft are provided with standard separation, the VMC minima feature visibility limits only, whereas in classes C–G airspace where some or all aircraft are not separated from each other by Air Traffic Control, the VMC minima also feature distance from cloud minima.

It is important not to confuse IMC with IFR (Instrument flight rules) – IMC describes the actual weather conditions, while IFR describes the rules under which the aircraft is flying. Aircraft can (and often do) fly IFR in clear weather, for operational reasons, or when flying in airspace where flight under VFR is not permitted; indeed by far the majority of commercial flights are operated solely under IFR. Any time a flight is operating in VMC, the crew are responsible for seeing and avoiding other traffic, however, since the flight is conducted under Instrument Flight Rules, ATC still provides separation services.

It is possible to be flying VFR in conditions that are legally considered VMC and have to rely on flight instruments for attitude control. Two examples would be on a dark night over

water (black hole effect), or a clear night with lights on the water and stars in the sky looking the same, i.e. no distinct horizon to fly by. (775words)

NEW WORDS

criteria	/kraɪˈtɪərɪə/	n.	标准
buffer	/ˈbʌfə/	n.	缓冲
confuse	/kənˈfjuːz/	v.	使……迷惑

NOTES

In Class A airspace, all operations must be conducted under Instrument Flight Rules (IFR) or Special Visual Flight Rules (SVFR) and are subject to ATC clearance. All flights are separated from each other by ATC.

Keys to Exercises

Lesson 1

IV. Vocabulary

1. transport	2. in-flight	3. airplanes	4. functioning	5. flight
6. while	7. turns	8. communicate	9. optimal	10. regulations

VI. Reading for More

1. A   2. C   3. A   4. B   5. D   6. A   7. C   8. C   9. D   10. B

Lesson 2

IV. Vocabulary

1. status	2. aviation	3. tasks	4. impact	5. Lacking
6. human	7. decisions	8. accurate	9. decision-making	10. emergency

VI. Reading for More

1. B   2. D   3. A   4. C   5. B   6. B   7. D   8. C   9. A   10. D

Lesson 3

IV. Vocabulary

1. training	2. cause	3. maneuvers	4. excessive	5. varies
6. stall	7. airplanes	8. circumstance	9. career	10. time

VI. Reading for More

1. B   2. C   3. B   4. A   5. D   6. A   7. B   8. C   9. B   10. C

Lesson 4

IV. Vocabulary for More

1. position	2. wing	3. extended	4. aileron	5. rudder
6. attitude	7. flap	8. stall	9. direction	10. excessively

VI. Reading for More

1. C   2. D   3. B   4. C   5. A   6. D   7. A   8. B   9. D   10. C

Lesson 5

IV. Vocabulary

1. warnings   2. recycling   3. landing   4. runways   5. installation
6. exterior   7. models   8. flat   9. well   10. proximity

VI. Reading for More

1.C   2. A   3. B   4.D   5. C   6.B   7. A   8. D   9.A   10. D

Lesson 6

IV. Vocabulary

1. displays   2. horizon   3. mounted   4. gyro   5. relative
6. align   7. cruising   8. exceeded   9. banked   10. scale

VI. Reading for More

1.D   2.C   3.A   4.D   5.C   6. D   7.D   8.C   9.B   10.B

Lesson 7

IV. Vocabulary

1. encountered   2. accelerate   3. malfunction   4. reject   5. overshooting
6. single   7. checklist   8. retract   9. cutoff   10. sufficient

VI. Reading for More

1. D   2. C   3. A   4. A   5. B   6. D   7. D   8. B   9. A   10. A

Lesson 8

IV. Vocabulary

1. cabin altitude   2. don   3. limiting   4. memory   5. recommended
6. idle   7. initiate   8. target   9. notify   10. safe altitude

VI. Reading for More

1. C   2. B   3. A   4. D   5.C   6. B   7. A   8.B   9. C   10.B

Lesson 9

IV. Vocabulary

1. transmitter   2. frequency   3. instrument   4. course   5. select

6. flying   7. relative   8. needle   9. indicate   10. followed

VI. Reading for More

1. A   2. D   3. D   4. B   5. C   6. B   7. C   8. D   9. B   10. B

Lesson 10

IV. Vocabulary

1. cockpit   2. available   3. propeller   4. alert   5. battery
6. landing   7. place   8. airflow   9. blinding   10. detected

VI. Reading for More

1. B   2. B   3. A   4. C   5. B   6. D   7. A   8. C   9. C   10. D

Lesson 11

IV. Vocabulary

1. end   2. rapidly   3. result   4. configuration   5. damage
6. airframe   7. accident   8. repair   9. crashed   10. tail

VI. Reading for More

1. D   2. C   3. A   4. C   5. A   6. A   7. D   8. B   9. D   10. C

Lesson 12

IV. Vocabulary

1. traffic   2. maneuvering   3. bank   4. factor   5. elevator
6. drag   7. performance   8. stall   9. relative   10. adjustment

VI. Reading for More

1. A   2. A   3. C   4. B   5. B   6. B   7. A   8. B   9. D   10. C

Lesson 13

IV. Vocabulary

1. operation   2. diversion   3. technical   4. flyable   5. assistance
6. location   7. abnormal   8. hydraulics   9. injury   10. sink

VI. Reading for More

1. D   2. C   3. C   4. A   5. B   6. C   7. B   8. D   9. C   10. D

Lesson 14

IV. Vocabulary

1. vehicle　　2. collision　　3. areas　　4. require　　5. operations

6. movement　　7. tower　　8. incidents　　9. standard　　10. awareness

VI. Reading for More

1. C　2. B　3. A　4. B　5. D　6. B　7. C　8. B　9. A　10. B

Lesson 15

IV. Vocabulary

1. issues　　2. route　　3. airports　　4. releases　　5. departures

6. ensure　　7. slot　　8. runway　　9. congested　　10. weather

VI. Reading for More

1. B　2. A　3. C　4. D　5. B　6. D　7. A　8. C　9. C　10. D

Lesson 16

IV. Vocabulary

1. clear　　2. streams　　3. moderate　　4.extreme　　5. control

6. inside　　7. unsecured　　8. foreseen　　9. aft　　10. enter

VI. Reading for More

1.D　2. C　3.A　4. B　5. C　6. D　7.C　8. C　9.C　10. B

Appendix I

Glossary

A

abnormal	/æb'nɔːməl/	adj.	非正常的	L4
abort	/ə'bɔːt/	v.	中断	L12
abruptly	/əb'rʌptli/	adv.	突然地	L5
accelerate	/æk'seləreit/	v.	加速	L5
accelerometer	/æk,selə'rɒmitə/	n.	加速计	L6
access	/'ækses/	n.	入口；进入	L14
accident	/'æksidənt/	n.	事故	L4
accumulation	/ə,kjuːmju'leiʃən/	n.	积累	L2
accumulator	/ə'kjuːmjuleitə/	n.	蓄压器	L5
accuracy	/'ækjurəsi/	n.	精确度	L6
acrobatic	/,ækrə'bætik/	adj.	表演的	L10
acronym	/'ækrənim/	n.	缩略词	L2
activities	/æk'tivitiz/	n.	活动 (pl.)	L2
actuation	/,æktʃu'eiʃən/	n.	启动，作动	L3
actuator	/'æktʃueitə/	n.	作动筒	L3
administrator	/əd'ministreitə/	n.	管理人	L1
advisory	/əd'vaisəri/	adj.	咨询的	L15
aerial	/'eəriəl/	adj.	航空的	L2
aerobatics	/,eərəʊ'bætiks/	n.	特技飞行	L3
aerodrome	/'eərə,drəum/	n.	机场	L5
aerodynamic	/eərədai'næmik/	adj.	空气动力的	L4
aerofoil	/'eərəfɔil/	n.	翼面	L3
aeronautical	/,eərə'nɔːtikəl/	adj.	航空的	L2
aggressively	/ə'gresivli/	adv.	攻击性地 过量地	L5
aileron	/'eilərɒn/	n.	副翼	L3
aircraft	/'eəkraːft/	n.	飞机；航空器	L1
airfield	/'eəfiːld/	n.	机场	L9

airframe	/'eə,freim/	*n.*	机体	L4
airline	/'eəlain/	*n.*	航空公司；航线	L1
airliner	/'eə,lainə/	*n.*	公司班机	L1
airmass	/'eəmæs/	*n.*	气团	L16
airspace	/'eə,speis/	*n.*	空域	L9
airspeed	/'eə'spi:d/	*n.*	空速	L2
airtight	/'eətait/	*adj.*	气密的	L8
airway	/'eəwei/	*n.*	航路	L6
alert	/ə'lə:t/	*n. & v.*	警告，警觉	L2
align	/ə'lain/	*v.*	对准	L12
allocate	/'æləkeit/	*v.*	分配	L9
allocation	/,ælə'keiʃən/	*n.*	分配	L15
allure	/ə'ljuə/	*v.*	诱惑	L14
aloft	/ə'lɒft/	*adj.*	高空的	L2
alteration	/,ɔ:ltə'reiʃən/	*n.*	更改	L10
alternate	/ɔ:l'tə:nit/	*adj.*	备用的	L5
	/ɔ:l'təneit/	*v.*	交替	
		n.	备降场	
alternative	/ɔ:l'tə:nətiv/	*adj.*	备用的,备选的	L13
alternator	/'ɔ:ltəneitə/	*n.*	交流机	L7
altitude	/'ælti,tju:d/	*n.*	高度	L2
amateur	/'æmətə:/	*adj.*	业余的	L10
amber	/'æmbə/	*n.*	琥珀色	L13
ambient	/'æmbiənt/	*a.*	外界的	L8
ambiguity	/,æmbi'gju:iti/	*n.*	模糊	L13
analog	/'ænələg/	*adj.*	模拟式的	L2
angle	/'ængl/	*n.*	角度	L3
annunciator	/ə'nʌnʃieitə/	*n.*	信号器	L7
antenna	/æn'tenə/	*n.*	天线	L12
anticipate	/æn'tisipeit/	*v.*	预期	L1
anvil	/'ænvil/	*n.*	砧状云	L16
approach	/ə'prəutʃ/	*n.& v.*	进近	L3
armed	/ɑ:md/	*adj.*	预位的	L10
armrest	/'ɑ:mrest/	*n.*	扶手	L1
array	/ə'rei/	*n.*	系列	L12
arrival	/ə'raivəl/	*n.*	进场；到达	L15

aspect	/ˈæspekt/	n.	方面	L2
assembly	/əˈsembli/	n.	装置	L6
assessment	/əˈsesment/	n.	评价，评估	L2
assign	/əˈsain/	n.	分配；指定	L15
associate	/əˈsəuʃieit/	v.	联系（with）	L2
asymmetric	/ˌeisiˈmetrikl/	adj.	不对称的	L4
audible	/ˈɔːdəbl/	adj.	音频的；可听见的	L10
audio	/ˈɔːdiəu/	n.	音频	L2
auditory	/ˈɔːditəri/	adj.	声音的，听觉的	L2
augment	/ɔːgˈment/	v.	增加，增益	L2
augmentation	/ˌɔːgmenˈteiʃən/	n.	增加；增益	L6
authority	/əˈθɔriti/	n.	权威，权限	L1
autopilot	/ˌɔːtəˈpailət/	n.	自动驾驶仪	L2
autothrottle	/ˌɔːtəˈθrɔtl/	n.	自动油门	L6
aviation	/ˌeiviˈeiʃən/	n.	航空	L2
avionics	/ˌeiviˈɒniks/	n.	航空电子	L6
axis	/ˈæksis/	n.	轴	L6

B

backup	/ˈbækˌʌp/	adj.	备用的，辅助的	L4
baggage	/ˈbægidʒ/	n.	行李	L1
balance	/ˈbæləns/	n.	平衡	L2
bank	/bæŋk/	n.	坡度	L4
bar	/bɑː/	n.	指令杆	L6
barotrauma	/ˌbærəˈtrɔːmə/	n.	耳气压伤	L8
base	/beis/	n.	四边；基地	L3
battery	/ˈbætəri/	n.	电瓶	L7
belly	/ˈbeli/	n.	机腹	L7
blast	/ˈblɑːst/	n.	一股(强烈的气流)	L14
blip	/blip/	n.	（雷达）物体光点	L9
boarding	/ˈbɔːdiŋ/	n.	登机	L1
booster	/ˈbuːstə/	n.	增压器	L7
bounce	/bauns/	v.	弹跳	L9
bracket	/ˈbrækit/	n.	支架	L10
brake	/breik/	n.	刹车	L5
breakdown	/ˈbreikˌdaun/	n.	故障	L5

breakout	/'breikaʊt/	n.	断开	L3
briefing	/ˈbriːfiŋ/	n.	讲述；讲评	L1
buffer	/'bʌfə/	n.	缓冲	L16
build-up	/'bild‚ʌp/	n.	雷雨	L16
buoyancy	/'bɔiənsi/	n.	浮力	L10
bus	/bʌs/	n.	汇流条	L7
bustle	/bʌsl/	n.	垫子	L10
buzzer	/'bʌzə/	n.	蜂鸣器	L13
bypass	/'bai‚pɑːs/	n.&v.	旁通；绕开	L5

C

cabinet	/'kæbinit/	n.	橱柜	L10
cable	/keibl/	n.	缆索，电缆	L3
calibrate	/'kæli‚breit/	v.	修正，校准	L2
camber	/'kæmbə/	n.	曲度	L11
canopy	/'kænəpi/	n.	舱盖	L10
capability	/‚keipə'biliti/	n.	能力	L15
capacity	/kə'pæsiti/	n.	容量	L1
captain	/'kæptin/	n.	机长	L1
cater	/'keitə/	v.	配餐；一致	L14
ceiling	/'siːliŋ/	n.	云高；升限	L5
cell	/sel/	n.	雷暴单体	L16
certification	/‚səːtifi'keiʃən/	n.	认证，证实	L4
certify	/'səːtifai/	v.	认证；认可；证明	L1
channel	/tʃænəl/	n.	频道；信道	L4
charge	/tʃɑːdʒ/	v.	充压，充电	L5
checklist	/'tʃeklist/	n.	检查单	L1
circuit	/'səːkit/	n.	电路；环路	L4
circuitry	/'səːkitri/	n.	电路	L4
circumnavigate	/‚səːkəm'nævigeit/	v.	绕航，绕飞	L16
clearance	/'kliərəns/	n.	许可；间距	L15
coating	/'kəʊtiŋ/	n.	涂层	L2
code	/kəʊd/	n.	编码	L9
coincide	/‚kɔin'said/	v.	巧合；一致	L9
collision	/kə'liʒən/	n.	相撞，撞机	L4
combustor	/kəm'bʌstə/	n.	燃烧室	L7

command	/kəˈmɑːnd /	n. & v.	指挥；指令	L1
communicate	/kəˈmjuːnikeit /	v.	交流；通讯	L1
compartment	/kəmˈpɑːtmənt/	n.	舱	L2
compass	/ˈkʌmpəs/	n.	罗盘	L9
compensate	/ˈkɒmpenseit/	v.	补偿	L4
complicate	/ˈkɒmplikeit/	v.	使……复杂化	L15
component	/kəmˈpəunənt/	n.	部件；矢量	L2
composite	/kəmˈpɒzit/	adj.	复合的	L9
concept	/ˈkɒnsept/	n.	概念	L15
concourse	/ˈkɒŋkɔːs/	n.	大厅	L14
condense	/kənˈdens/	v.	凝结	L16
configuration	/ˌkənfigəˈreiʃən/	n.	形态，外形	L3
confirm	/kənˈfəːm/	v.	证实；确认	L5
confuse	/kənˈfjuːz/	v.	使……迷惑	L16
confusion	/kənˈfjuːʒən/	n.	迷惑，困惑	L14
consolidate	/kənˈsɒlideit/	v.	巩固；加强	L15
contact	/ˈkɒntækt/	n.	接触；联系	L5
container	/kənˈteinə/	n.	容器	L5
contaminant	/kənˈtæminənt/	n.	污染物	L7
contaminated	/kənˈtæmineitid/	adj.	被污染的	L4
controllability	/ˌkəntrəuləˈbiliti/	n.	操纵性	L5
convection	/kənˈvekʃən/	n.	对流	L16
convert	/kənˈvəːt/	v.	转换	L4
convey	/kənˈvei/	v.	传递	L9
co-ordination	/kəuˈɔːdiˈneiʃən/	n.	协调	L15
counter-terrorism	/ˈkauntəˈterərizm/	n.	反恐怖	L14
couple	/ˈkʌpl/	v.	耦合	L5
course	/kɔːs/	n.	航道	L2
coverage	/ˈkʌvəridʒ/	n.	覆盖范围	L15
crab	/kræb/	n.	侧航	L11
crank	/kræŋk/	n.	曲轴	L4
crash	/kræʃ/	v.	坠毁	L5
crew	/kruː/	n.	人员，组员	L1
criteria	/kraiˈtiəriə/	n.	标准	L16
crosscheck	/ˈkrɒstʃek/	n.	交叉检查	L6
crosswind	/ˈkrɒswind/	n.	侧风；二边	L5

crystal	/ˈkristəl/	n.	晶体	L16
cue	/kjuː/	n.	提示；征候	L12
cumulonimbus	/ˌkjuːmjuləʊˈnimbəs/	n.	积雨云	L16
cumulus	/ˈkjuːmjuləs/	n.	积云	L16
curl	/kəːl/	v.	卷曲	L3
current	/ˈkʌrənt/	n.	电流	L7

D

damper	/ˈdæmpə/	n.	阻尼器	L6
database	/ˈdeitə,beis/	n.	数据库	L6
datalink	/ˈdeitəlink/	n.	数据联路	L6
deactivate	/diˈæktiveit/	v.	断开	L1
debris	/ˈdebriː/	n.	碎片	L4
decay	/diˈkei/	v.	消减	L5
deck	/dek/	n.	甲板	L2
demonstration	/ˌdemənˈstreiʃən/	n.	演示	L1
density	/ˈdensiti/	n.	密度	L2
depart	/diˈpɑːt/	v.	起飞，离场	L2
deploy	/diˈplɔi/	v.	放出，张开	L11
depressurisation	/ˌdipreʃəraiˈzeiʃən/	n.	释压	L1
descent	/diˈsent/	n.	下降	L2
designator	/ˌdezigˈneitə/	n.	号，指示器	L14
destruction	/diˈstrʌkʃən/	n.	破坏	L10
detach	/diˈtætʃ/	v.	分开；放下	L10
detect	/diˈtekt/	v.	探测	L6
deterioration	/diˌtiriəˈreiʃən/	n.	恶化，变差	L4
detour	/ˈdiːtʊə/	v.	绕航，绕飞	L16
deviate	/ˈdiːvieit/	v.	偏离(from)	L1
deviation	/ˌdiːviˈeiʃən/	n.	偏离	L2
dew-point	/ˈdjuːpɔint/	n.	露点	L16
dial	/ˈdaiəl/	n.	仪表盘	L2
diameter	/daiˈæmitə/	n.	直径	L6
differential	/ˌdifəˈrenʃəl/	n.	差别	L3
digital	/ˈdidʒitəl/	adj.	数字式的	L3
diligence	/ˈdilidʒəns/	n.	勤奋	L10
disarm	/disˈɑːm/	v.	解除预位	L1

discomfort	/dis'kʌmfət/	n.	不舒适	L4
discrete	/dis'kriːt/	adj.	分离的，分开的	L9
discretion	/dis'kreʃən/	n.	决定；谨慎	L1 L9
disembark	/ˌdisim'bɑːk/	v.	下飞机	L1/L14
disintegrate	/dis'intigreit/	v.	解体；瓦解	L13
dis-orientation	/disˌɔːrien'teiʃən/	n.	迷失方向；迷航	L4
dispatcher	/dis'pætʃə/	n.	签派员	L6
displace	/dis'pleis/	v.	位移	L6
disposition	/ˌdispə'ziʃən/	n.	处置，处理	L1
distract	/di'strækt/	v.	分心；分散注意力	L13
ditch	/ditʃ/	v.	壕沟；水上迫降	L5
dive	/daiv/	v.	俯冲	L5
divert	/dai'vəːt/	v.	改航；备降	L12
downstream	/daun'striːm/	n.	下游	L3
downwind	/daun'wind/	n.	顺风；三边	L3
drift	/drift/	v.& n.	偏流；偏移	L3
droplet	/'drɒplit/	n.	水滴	L16
drunken	/'drʌkən/	adj.	醉酒的	L1
dual	/'djuəl/	adj.	双重的；双的	L6/L7
duct	/dʌkt/	n.	管道	L8
duplicate	/'duːplikeit/	a.	复制的；双重的	L8

E

electrical	/i'lektrikəl/	adj.	电的；电气的	L4
electronic	/iˌlek'trɔnik/	adj.	电子的	L2
elevator	/'eliveitə/	n.	升降舵	L3
eliminate	/i'limineit/	v.	消除	L15
embed	/im'bed/	v.	嵌入	L16
emergency	/i'məːdʒənsi/	n.	紧急	L1
emission	/i'miʃən/	n.	释放	L11
enclose	/in'kləuz/	v.	密闭	L2
encounter	/in'kauntə/	v.	遭遇	L4
engage	/in'geidʒ/	v.	接通	L6
ensure	/in'ʃuə/	v.	确保；保证	L2
envelope	/'enviləup/	n.	包线	L4
equivalent	/i'kwivələnt/	adj.	相等的；等量的	L3

erectable	/i'rektəbl/	adj.	可直立的	L10
error	/'eərə/	n.	错误；误差	L15
escape	/i'skeip/	n.	逃离	L10
escort	/'eskɔ:t/	v.	护送	L1
evacuation	/i:ˌvækju'eiʃən/	n.	撤离	L1
evolve	/i'vɔlv/	v.	发展；进化	L15
excess	/ik'ses/	n.	过多	L5
execute	/'eksikju:t/	v.	执行	L5
exhaustion	/ˌig'zɔ:stʃən/	n.	耗尽	L2
exit	/'eksit/	n. & v.	出口；出去	L1/L14
expel	/ˌiks'pel/	v.	挤出	L5
explosive	/iks'pləusiv/	n.	爆炸性物品	L14
extend	/iks'tend/	v.	伸出、放出	L3

F

fasten	/'fɑ:sn/	v.	系	L1
fatal	/'feitəl/	adj.	致命的	L4
faulty	/fɔ:lti/	adj.	故障的	L5
feasible	/'fi:zəbl/	adj.	可行的	L15
feature	/'fitʃə/	n.	装置	L9
features	/'fi:tʃəz/	n.	装置	L1
feedback	/'fi:dbæk/	n.	反馈	L6
feeder	/'fi:də/	n.	支线	L12
filter	/'filtə/	n.	过滤器	L5
fin	/fin/	n.	立尾	L4
final	/'fainəl/	n.	五边	L3
		adj.	最后的	
flap	/flæp/	n.	襟翼	L3
flare	/fleə/	n.& v.	拉平	L3/L10
flashlight	/'flæʃlait/	n.	手电筒	L1
flip	/flip/	v.	翻转	L13
flow	/fləu/	n.	流量	L15
fluid	/flu:id/	n.	液体；液压油	L4
foam	/fəum/	n.	泡沫；铺设泡沫	L5
forecast	/fɔkɑ:st/	v.	预报	L9
format	/'fɔ:mæt/	n.	格式；形势	L15

freight	/freit/	n.	货物	L14
friction	/'frikʃən/	n.	摩擦	L3
front	/frʌnt/	n.	锋面	L16
fuel	/fjuəl/	n.	燃油	L1
fuelling	/'fjuəliŋ/	n.	加油	L1
fume	/fju:m/	n.	烟雾	L4
fuselage	/'fju:zilidʒ/	n.	机身	L4

G

galley	/'gæli/	n.	厨房	L8
gap	/gæp/	n.	间歇；间隔	L12
garbage	/'gɑ:bidʒ/	n.	垃圾	L1
gate	/geit/	n.	登机门	L14
gearbox	/'giəbɒks/	n.	变速箱；齿轮箱	L7
generate	/'dʒenəreit/	v.	发电	L7
generator	/'dʒenəreitə/	n.	发电机	L7
glareshield	/'gleə,ʃi:ld/	n.	遮光板	L2
groove	/gru:v/	n.	槽纹	L14
grossly	/'grəusli/	adv.	过多地	L10
guidance	/'gaidəns/	n.	制导，引导	L12
gusty	/'gʌsti/	adj.	阵风的	L16
gyro	/'dʒaiərəu/	n.	陀螺	L6

H

hand-over	/hænd'əuvə/	n.	移交	L5
hangar	/'hæŋə/	n.	机库	L14
hastily	/'heistili/	adv.	匆忙地；草率地	L2
hazard	/'hæzəd/	n.	危险	L3
hazardous	/'hæzədəs/	adj.	危险的	L2
headset	/'hedset/	n.	耳机	L8
heal	/hi:l/	v.	治愈	L8
heater	/'hi:tə/	n.	加温器	L8
helicopter	/'helikɒptə/	n.	直升机	L1
hijack	/'hai,dʒæk/	n.	劫机	L9
horizon	/hə'raizən/	n.	地平线	L4

horizontal	/ˌhɒriˈzɒntl/	adj.	水平的	L2
horn	/hɔːn/	n.	喇叭	L13
humidity	/hjuːˈmiditi/	n.	湿度	L8
hydraulic	/haiˈdrɔːlik/	adj.	液压的	L3
hydro-clean	/ˌhaidrəˈkliːn/	v.	洁水	L14
hydro-mechanical	/haidrəˌmiˈkænikəl/	adj.	液压机械的	L4
hydroplane	/ˈhaidrəplein/	v.	滑水	L14
hypoxia	/haiˈpɒksiə/	n.	缺氧	L8

I

icing	/ˈaisiŋ/	adj.	结冰	L4
ident	/ˈaidənt/	n.	识别、识别符	L4
idle	/ˈaidl/	adj.&v.	慢车	L7
ignite	/igˈnait/	v.	点燃	L5
illuminate	/iˈljuːmineit/	v.	闪烁，亮	L7
imminent	/ˈiminənt/	adj.	立刻的，迫近的	L4
impact	/ˈimpækt/	n.	撞击；影响	L4
impair	/imˈpeə/	v.	损害	L8
inaccuracy	/inˈækjurəsi/	n.	不准确	L10
inboard	/ˈinbɔːd/	adj.	内侧	L3
incapacitation	/ˈinkəˌpæsiˈteiʃn/	n.	失能	L1
inclusive	/inˈklusiv/	adj.	全部的	L9
increment	/inˈkrimənt/	n.	增量	L9
incur	/inˈkə/	v.	导致	L6
induce	/inˈdjuːs/	v.	诱发	L5
inevitable	/inˈevitəbl/	adj.	必然的	L5
infection	/inˈfekʃən/	n.	感染；传染	L8
inflate	/inˈfleit/	v.	充气	L10
infringe	/inˈfrindʒ/	v.	侵犯	L8
initiate	/iˈniʃieit/	v.	开始；启用	L6
input	/ˈinput/	n.& v.	输入	L6
instability	/ˌinstəˈbiliti/	n.	不稳定性	L16
instrument	/ˈinstrumənt/	n.	仪表	L2
insulation	/ˌinsjuˈleiʃən/	n.	绝缘	L7
integrity	/inˈtegriti/	n.	统一	L2
intensity	/inˈtensiti/	n.	强度	L3

intention	/ˌin'tenʃən/	n.	意图	L5
interaction	/ˌintə'ækʃən/	n.	互动；交流	L2
interfere	/ˌintə'fiə/	v.	干扰	L4
intermediate	/ˌintə'mi:diət/	adj.	中间的	L12
interpose	/ˌintə'pəuz/	v.	介于……之间	L4
interrogation	/inˌterə'geiʃən/	n.	询问	L9
intervention	/ˌintə'venʃən/	n.	接入，干预	L4
inward	/'inwəd/	adj.	内侧的	L3

J

jam	/dʒæm/	n.& v.	卡阻	L3
jet	/dʒet/	n.	喷气机	L12
justification	/ˌdʒʌstifi'keiʃn/	n.	证明	L1

K

keyboard	/'ki:ˌbɔ:d/	n.	键盘	L6
keystroke	/'ki:strəuk/	n.	敲击键盘	L6
knob	/knɒb/	n.	旋钮	L2

L

landing	/'lændiŋ/	n.	着陆	L2
lane	/lein/	n.	道	L10
lanyard	/'lænjəd/	n.	牵索	L10
lapse	/'læps/	n.	流逝（时间）	L7
laser	/'leizə/	n.	激光	L6
lavatory	/'lævətəri/	n.	洗手间	L1
leak	/li:k/	n.	泄漏	L4
leeward	/'li:wəd/	adj.	向下风的，背风的	L11
leg	/leg/	n.	段；边；支柱	L5
lethality	/li'θæliti/	n.	致命	L14
lever	/'levə/	n.	手柄	L6
license	/'laisəns/	n. & v.	执照；授证	L1
lifevest	/'laifvest/	n.	救生衣	L1
lightning	/'laitniŋ/	n.	闪电	L16
link	/link/	n. & v.	连接，联路	L3
load	/ləud/	n.	负载；负荷	L1

		v.	装载	
logic	/'lɒdʒik/	n.	逻辑	L6
loudspeaker	/laʊd'spi:kə/	n.	扬声器	L8
lubrication	/ˌlu:bri'keiʃən/	n.	润滑	L4

M

Mach	/mɑːk/	n.	马赫数	L3
macro-texture	/ˌmækrə'teksitʃə/	v.	高质地化	L14
magenta	/mə'dʒentə/	n.	洋红色	L6
maintenance	/'meintinəns/	n.	维护，维修	L4
malfunction	/mæl'fʌŋkʃən/	n.	故障	L4
mandatory	/'mændətəri/	adj.	强制的	L9
maneuver	/mə'nu:və/	n.	机动	L6
maneuverability	/mə,nuvərə'bilti/	n.	机动性	L3
manipulate	/mə'nipjuleit/	v.	操纵	L1
manual	/'mænjuəl/	adj.	人工的	L3
		n.	手册	
margin	/'mɑːdʒin/	n.	余度	L11
marginal	/'mɑːdʒinəl/	adj.	边缘的	L4
marking	/'mɑːkiŋ/	n.	标线	L14
mast	/mæst/	n.	杆	L9
mechanic	/mi'kænik/	n.	机务员，机械人员	L1
mechanical	/mi'kænikəl/	adj.	机械的	L1
mechanism	/'mekənizm/	n.	机构，机械装置	L3
medium	/'mi:diəm/	adj.	中等的	L14
memo	/'memə/	n.	备忘，记忆	L8
merge	/mə'dʒ/	v.	兼并，并入	L14
meteorological	/ˌmitiərə'lɒdʒikəl/	adj.	气象的	L9
microburst	/'maikrə,bə:st/	n.	微下击气流	L4
microphone	/'maikrəfəun/	n.	麦克风；话筒	L8
minima	/'minimə/	n.	最低值；最低标准	L12
minuscule	/'minəskjul/	adj.	微小的	L4
mitigate	/'miti,geit/	v.	减小；减轻	L4
modification	/ˌmɒdifi'keiʃən/	n.	修改	L2
modify	/'mɒdifai/	v.	修改	L4
modulation	/ˌmɒdju'leiʃən/	n.	调制；调解；调节	L12

moisture	/ˈmɔistʃə/	n.	水分	L16
monetary	/ˈmʌnəteri/	adj.	货币的；金融的	L15
monitor	/ˈmɒnitə/	v.	监控	L1
moor	/muə/	v.	抛锚	L10
motion	/ˈməuʃən/	n.	运动	L3
motorcycle	/ˈməutə,saikl/	n.	摩托车	L2
multiple	/ˈmʌltipl/	adj.	多个，多重的	L4
multistory	/ˌmʌltiˈstɔːri/	n.	多层	L14

<h1 style="text-align:center">N</h1>

navigation	/ˌnæviˈgeiʃən/	n.	导航；领航	L1
needle	/ˈniːdl/	n.	指针	L12
negative	/ˈnegətiv/	adj.	负的；否定的	L4
neutral	/ˈnjuːtrəl/	adj.	中立（位）的	L6
neutralize	/ˈnjuːtrə,laiz/	v.	中立	L2
nocturnal	/nɒkˈtəːnəl/	adj.	夜间的，晚上的	L16
nominal	/ˈnɒminəl/	adj.	额定的	L11
notoriety	/ˌnəutəˈraiəti/	n.	恶名	L8
nullify	/ˈnʌlifai/	v.	使无效	L10

<h1 style="text-align:center">O</h1>

obstacle	/ˈɒbsteikl/	n.	障碍物	L2
occupant	/ˈɒkjupənt/	n.	乘员	L4
omission	/əˈmiʃən/	n.	省略	L15
omni-bearing	/ˌɒmniˈbeəriŋ/	n.	全方位	L12
operation	/ˌɒpəˈreiʃən/	n.	运转，工作，操作	L1
operator	/ˈɒpə,reitə/	n.	操作人员	L4
optimize	/ˈɒptimaiz/	v.	使……最佳	L15
oscillation	/ˌɒsiˈleiʃən/	n.	震荡	L4
outboard	/ˈautbɔːd/	adj.	外侧	L3
outflow	/ˈautfləu/	n.	外流	L8
outlet	/ˈautlet/	n.	出口	L5
outward	/ˈautwəd/	adj.	外侧的	L3
overboard	/ˈəuvəbɔːd/	adv.	舱外	L8
overboost	/ˈəuvəˈbuːst/	n.	过度增压	L7
overlap	/ˈəuvəˈlæp/	v.	重叠	L9

overload	/ˌəuvəˈləud/	n.	过载	L2
override	/ˌəuvəˈraid/	v.	超控	L1
oxygen	/ˈɒksidʒən/	n.	氧气	L8

P

pack	/pæk/	n.	空调组件	L8
pad	/pæd/	n.	平台；垫	L14
paddle	/ˈpædl/	n.	闸门，折动板	L10
panel	/ˈpænəl/	n.	面板	L2
parachute	/ˈpærəʃuːt/	n.	降落伞	L6
parameter	/pəˈræmitə/	n.	参数	L4
partial	/ˈpɑːʃəl/	adj.	部分的	L5
particle	/ˈpɑːtikl/	n.	颗粒	L5
path	/pɑːθ/	n.	轨迹	L6
pavement	/ˈpeivmənt/	n.	道面	L14
peculiar	/piˈkjuːljə/	n.	特殊的	L11
pedal	/ˈpedəl/	n.	脚蹬，踏板	L3
pedestal	/ˈpedistl/	n.	操纵台	L2
penetrate	/ˈpenitreit/	v.	穿越	L16
performance	/pəˈfɔːməns/	n.	性能	L2
perishable	/ˈpeːriʃəbl/	adj.	易腐烂的	L14
phase	/feiz/	n.	阶段	L2
physiological	/ˌfiziəˈlɒdʒikəl/	adj.	生理的	L8
pipe	/paip/	n.	管道	L4
piston	/ˈpistən/	n.	活塞	L5
pitch	/pitʃ/	n.	俯仰	L3
pivot	/ˈpivət/	n.	枢轴	L11
		v.	绕着枢轴转动	
placard	/ˈplækɑːd/	n.	标牌	L10
plate	/pleit/	n.	图	L12
plot	/plɒt/	v.	标出	L9
pneumatic	/njuːˈmætik/	a.	气源的	L8
pneumothorax	/ˌnjuːməˈθɔːræks/	n.	气胸	L8
positive	/ˈpɒzitiv/	adj.	正的，积极的	L6
potential	/pəˈtenʃəl/	n.	潜力；可能	L5
		adj.	潜在的；可能的	

precarious	/pri'keəriəs/	a.	危险的	L11
precaution	/pri'kɔːʃən/	n.	预防	L1
precipitation	/pri,sipi'teiʃən/	n.	降水	L16
precision	/pri'siʒən/	n.	精密，精确	L12
precool	/pri'kuːl/	v.	预制冷	L8
predetermine	/priːdi'təːmin/	v.	预定	L5
predominance	/pri'dɔminəns/	n.	主宰，支配	L12
preflight	/'pri'flait/	n.	飞行前	L2
pressurization	/,preʃərai'zeiʃən/	n.	增压	L4
procedure	/prə'siːdʒə/	n.	程序	L1
profile	/'prəʊfail/	n.	剖面	L7
progress	/'prəʊgres/	n.	进程	L2
prohibit	/prə'hibit/	v.	禁止	L15
prolong	/prə'lɒŋ/	v.	延长	L11
proportion	/prə'pɔːʃən/	n.	比例	L4
provision	/prə'viʒn/	n.	条款	L1
proximity	/prɒk'simiti/	n.	附近	L12
prudent	/'pruːdənt/	adj.	谨慎的	L7
pump	/pʌmp/	n.	泵	L5
purser	/'pəːsə/	n.	乘务长	L1
pushbutton	/'puʃ'bʌtən/	n.	按钮	L7

Q

quadruplex	/'kwɒdrupleks/	n.	四倍	L4
quantity	/'kwɒntiti/	n.	数量	L1
query	/'kweri/	v.	询问	L2

R

radar	/'reidə/	n.	雷达	L9
raft	/rɑːft/	n.	安全筏	L10
rate	/reit/	n.	比率	L4
rating	/'reitiŋ/	n.	等级	L2
readability	/,riːdə'biliti/	n.	清晰度	L8
recirculate	/ri'səːkjuleit/	v.	再循环	L8
recovery	/ri'kɒvəri/	n.	改出	L3
rectify	/'rektifai/	v.	纠正	L1

redundancy	/ri'dʌndənsi/	n.	余度，多余	L2
redundant	/ri'dʌnənt/	adj.	富余，多余的	L8
reference	/'refərəns/	n.	参考；基准	L12
reflection	/ri'flekʃən/	n.	折射；思考	L12
refrigerate	/ri'fridʒərit/	v.	制冷	L8
refuel	/ri'fjuəl/	v.	加油	L6
regime	/'redʒim/	n.	状态，方式	L4
regulate	/'regjuleit/	v.	调节	L8
reintroduce	/ˌriːintrə'djuːs/	v.	再流入	L8
relay	/ri'lei/	n.	继电器	L12
		v.	转递；转发	
release	/ri'liːs/	v.	松开；释放	L5
relieve	/ri'liːv/	v.	减轻	L6
replace	/ri'pleis/	v.	取代；更换	L2
replenish	/ri'pleniʃ/	v..	更新；补充	L5
rescue	/'reskju:/	n. & v.	营救	L5
reserve	/ri'zəːv/	n.	储备；备份	L5
reservoir	/'rezəvwɑː/	n.	(液压)油箱	L5
resistance	/ri'zistəns/	n.	阻力；阻滞	L5
respective	/ri'spektiv/	adj.	各自的；分别的	L15
restriction	/ri'strikʃən/	n.	限制	L15
retain	/ri'tein/	v.	保留	L8
retard	/ri'tɑːd/	v.	回收	L13
retract	/ri'trækt/	v.	收上	L5
retrieve	/ri'triːv/	v.	获得	L14
revenue	/'revinjuː/	n.	(财政)收入	L10
reversion	/ri'vəːʃən/	n.	恢复	L3
risk	/risk/	n. & vt.	风险	L2
rod	/rɒd/	n.	连杆	L3
rollout	/'rəʊl'aʊt/	n.	滑跑	L13
rotate	/rəʊ'teit/	v.	转动，抬轮	L3
rotation	/rəʊ'teiʃən/	n.	转动；抬轮	L11
rough	/rʌf/	a.	粗糙的	L5
route	/ruːt/	n.	航线	L6
routing	/'ruːtiŋ/	n.	管路	L4
rudder	/'rʌdə/	n.	方向舵	L3

runway	/ˈrʌnwei/	n.	跑道	L1

S

saddleback	/ˈsædlbæk/	n.	鞍形背	L16
saturated	/ˈsætʃəreitid/	adj.	饱和的	L7
scanning	/ˈskæniŋ/	n.	扫视	L2
scenario	/siˈnɑːriəu/	n.	情景；场景	L4
screen	/skriːn/	n. & v.	过滤；屏幕	L5
seal	/siːl/	v.	密封	L3
seamless	/ˈsiːmlis/	adj.	无缝的	L15
seatbelt	/ˈsiːtˈbelt/	n.	安全带	L1
security	/siˈkjuːriti/	n.	安全；保安	L14
self-contained	/ˈselfkənˈteind/	adj.	自带的	L7
seniority	/ˌsiːniˈɔːrəti/	n.	资历	L1
sensor	/ˈsensə/	n.	传感器	L4
separation	/ˌsepəˈreiʃən/	n.	间隔	L15
sequence	/ˈsiːkwəns/	n.	顺序	L5
		v.	调序	
servo	/ˈsəːvə/	n.	伺服机构	L6
shaft	/ʃɑːft/	n.	轴	L7
shoulder	/ˈʃəuldə/	n.	肩部	L8
sideward	/ˈsaidwəd/	adj.	侧向的	L13
sill	/sil/	n.	梁；槛	L10
sinus	/ˈsainəs/	n.	鼻窦	L8
slat	/slæt/	n.	缝翼	L3
slide	/slaid/	n.	滑梯	L10
slip	/slip/	n.	侧滑	L13
slope	/sləup/	n.	斜坡；坡度	L11
slot	/slɒt/	n.	缝	L3
smoothly	/ˈsmuːðli/	adv.	柔和地	L11
sophisticated	/səˈfistiˌkeitid/	adj.	复杂的	L5
spark	/spɑːk/	n.	火花	L5
spin	/spin/	n. & v.	螺旋	L4
spoiler	/ˈspɒilə/	n.	扰流板	L3
spring	/spriŋ/	n.	弹簧	L5
square	/skweə/	n.	平方	L15

squawk	/skwɔːk/	*n. & v.*	调定应答机	L9
squib	/skwib/	*n.*	爆管	L10
stability	/stə'bility/	*n.*	稳定性	L4
stabilizer	/'steibilaizə/	*n.*	安定面	L3
stage	/'steidʒ/	*n.*	级	L7
standby	/'stændbai/	*n.*	备用的	L3
standpipe	/'stænd,paip/	*n.*	竖管	L5
station	/'steiʃən/	*n.*	站位；电台	L1
status	/'steitəs/	*n.*	状态	L2
steerable	/'stiərəbl/	*adj.*	可操纵的	L11
stopway	/'stɒpwei/	*n.*	安全道	L14
stow	/stəʊ/	*v.*	收放	L1
stress	/stres/	*n.*	应力，压力	L10
strip	/strip/	*n.*	条；带	L5
structure	/'strʌktʃə/	*n.*	结构	L10
submerge	/səb'məːdʒ/	*v.*	浸没	L10
subset	/səb'set/	*n.*	子集；部分	L6
supervision	/,sjuːpə'viʒən/	*n.*	监管，管理	L4
supervisory	/,sjuːpə'vaizəri/	*adj.*	监管的，管理的	L12
suppress	/sə'pres/	*v.*	压制	L7
surveillance	/sə'veləns/	*n.*	监控	L9
sustain	/sə'stein/	*v.*	持续	L6
swirl	/swəːl/	*v.*	旋转	L3
switch	/switʃ/	*n. & v.*	开关；转换	L2

T

tailor	/'teilə/	*v.*	调整	L4
tailplane	/'teilplein/	*n.*	尾翼	L4
taxi	/'tæksi/	*n. & v.*	滑行	L1
terminal	/'təːminl/	*n.*	候机楼	L14
		adj.	终端的	
terrain	/'terein/	*n.*	地形	L2
thermal	/'θəːməl/	*adj.*	热的	L5
throttle	/'θrɒtl/	*n.*	油门	L2
thrust	/'θrʌst/	*n.*	推力	L6
thunderstorm	/'θʌndə,stɔːm/	*n.*	雷雨	L16

tornado	/tɔːˈneidəu/	*n.*	龙卷风	L16
torque	/tɔːk/	*n.*	扭力	L3
tow	/təu/	*v.*	牵引	L5
track	/træk/	*n. & v*	航迹；跟踪	L2
transfer	/trænsˈfə/	*v.*	转换	L3
transit	/ˈtrænsit/	*v. & n.*	过渡	L13
transition	/trænˈziʃən/	*n.*	过渡	L11
transmit	/trænsˈmit/	*v.*	发射	L9
transponder	/trænsˈpɒndə/	*n.*	应答机	L9
transportation	/ˌtrænspəˈteiʃən/	*n.*	交通；运输	L14
trim	/trim/	*n. & v.*	配平	L3
triplex	/ˈtripleks/	*n.*	三倍	L4
tropical	/ˈtrɒpikəl/	*adj.*	热带的	L1
tube	/tjuːb/	*n.*	筒、管	L3
tuck	/tʌk/	*v.*	收上	L6
tune	/tjuːn/	*v.*	调谐	L12
turbine	/ˈtəːbain/	*n.*	涡轮	L7
turbofan	/ˈtəːbəufæn/	*n.*	涡扇	L7
turbulence	/ˈtəːbjuləns/	*n.*	颠簸气流	L1

U

ultralight	/ˌʌltrəˈlait/	*n.*	超轻型飞机	L11
unit	/ˈjuːnit/	*n.*	组件	L5
unserviceable	/ˈʌnˈsəːvisəbl/	*adj.*	不工作的	L1
update	/ʌpˈdeit/	*n. & v.*	更新	L2
updraft	/ˈʌpdrʌpt/	*n.*	上升气流	L16
uplock	/ˈʌplɔk/	*n.*	上位锁	L13
upright	/ˈʌprait/	*adj.*	直立	L1
upset	/ʌpˈset/	*n.*	失去安定性	L4
upwind	/ˈʌpˈwind/	*n.*	逆风；一边	L3
utensil	/juːˈtensl/	*n.*	器皿	L14

V

vacation	/veiˈkeiʃən/	*n.*	脱离（跑道）	L12
validate	/ˈvælideit/	*v.*	使生效	L6
valve	/vælv/	*n.*	活门	L3

variable	/ˈveəriəbl/	n.	变量	L3
variation	/ˌveəriˈeiʃən/	n.	变化，偏差	L9
veer	/viə/	v.	改变	L5
ventilate	/ˈventileit/	v.	通风	L8
versatility	/ˌvəːsəˈtiliti/	n.	多功能，多能力	L2
vertical	/ˈvəːtikl/	adj.	垂直的	L2
vibration	/vaiˈbreiʃən/	n.	震动	L7
vicinity	/viˈsiniti/	n.	附近	L12
visibility	/ˌviziˈbiliti/	n.	能见度	L5
visual	/ˈvizjuəl/	adj.	目视的	L4
voltage	/ˈvɒltidʒ/	n.	电压	L7
vortice	/ˈvəːtis/	n.	涡系	L3

W

wake	/weik/	n.	尾流	L3
waypoint	/ˈweiˌpɔint/	n.	航路点	L6
wheel	/wiːl/	n.	机轮；盘	L5
windmilling	/ˈwindmiliŋ/	n.	风转；风磨	L7
windscreen	/ˈwindskriːn/	n.	风挡	L2
windward	/ˈwindwəd/	adj.	向风的；迎风的	L11
wing	/wiŋ/	n.	机翼	L1
wingspan	/ˈwiŋspæn/	n.	翼展	L3
wingtip	/ˈwiŋtip/	n.	翼尖	L3
wire	/waiə/	n.	电线	L4

Y

yaw	/jɔː/	n.	偏航	L2
yoke	/jəuk/	n.	操纵杆	L11

Appendix II

Eevents and Domains

The following inventory of events, domains and sub-domains are some that characterize the day-to-day communications of air traffic controllers and pilots. These "events" represent control situations, routine or non-routine, which all controllers must be able to handle. Each event may require familiarity with many lexical domains, to which related words are associated.

1. EVENTS, DOMAINS, AND SUBDOMAINS IN AERODROME CONTROL

Airmiss(es) Air traffic rules; avoiding action; trajectory/flight path; speed; distance/range; aircraft characteristics; position.

Airshows Traffic information; activity; acrobatics; formation flights; procedures.

Approach delays Holding instructions; holding procedures; aerodrome circuit; endurance; diversion/alternate; necessary conditions; CAT III; all-weather landings.

Belly landing Attempted manoeuvres; status of lights; visual check (low pass); position of landing gear; endurance; fuel remaining; fuel dumping/jettisoning; speed; traffic information; state of runway; aerodrome environment; airport installations; emergency evacuation; emergency slides/escape chutes, etc.; fire hazard/risk; damage; ground services.

Bird risk/hazard Position; quantity; names/types of birds; bird scaring in progress; damage to aircraft; delays; bird scaring methods; behaviour of birds.

Bomb threat/alert/scare Disembarking passengers; diversion; baggage identification; fuel dumping/jettisoning; aircraft interior; crew actions/behaviour; ground services; airport installations.

Cargo problems/ dangerous goods Customs; type of cargo; (perishable) organs for transplant; toxic substances; handling; packaging; veterinary services; police search; sniffer dogs; load badly fixed or damaged; intercepting; impounding.

Fire on board Ground services; aircraft interior; smoke; asphyxia; smells; oxygen masks; warning lights; firefighting equipment; extinguishers; injuries; burns; medical assistance; fire brigade/firemen; emergency slides/escape chutes; engine shutdown; evacuation.

Ground movement incidents Activity on the field; fire brigade training exercises and interventions; vehicles on the field; braking action and visibility; traffic information; startup;

towing equipment; engine checks; remote holding pattern; holding point; runway infringement; delays; stuck in the mud; damage caused by vehicles on the ground; no entry disregarded; collisions; vehicle or plane breakdown; damage to beacons; foreign objects (name, description); problems boarding or disembarking passengers; baggage identification; means of disembarking; health services; handicapped/sick passengers; parking position/space.

Health problems Symptoms; first aid; aircraft interior; type of medical assistance; medical background of passengers; diversion; airport installations; ground services; sickness; discomfort; wounds; epidemics; medical equipment; blood (group, transfusion, etc.); medical advice; the human body; forensic surgeon; quarantine; food poisoning; food; vaccines; medical staff; medicines and artificial limbs.

Incidents on landing Long/short landing; missed exit; stuck in mud; weather; cargo problems; runway confusion; bird or animal hazard; damage to tires; aircraft breakdown; missed approach.

Industrial action Ground staff; control/operational staff; effects on trafffic; delays; types of strike; demonstrations; sit-ins.

MET (weather) conditions ATIS (visibility, clouds, etc.); (thunder) storms, lightning; damage and breakdown; snow clearing; gusts; wind shear and microburst; minima; state of runway; tailwind, crosswind; braking action; runway visual range; temperature inversion; turbulence; natural disasters; runway closed; change of runway.

Missed approach Go-around; minima; traffic position; endurance; reasons; traffic; procedures; speed.

Parachute jumping/dropping activity Position; information on other traffic and activity; duration of drop; drop zone.

Pilot not familiar with airfield Procedures; airfield installations; ground services.

Pilot's temporary disability Health problems; aircraft controls and instruments; pilot's actions/behaviour; airfield environment; airport installations.

Problems linked to flight plan Delays; slots; flight plan updating; computer.

Problems linked to passenger's behaviour + unlawful inferference Violent/threatening behaviour; reasons (drunkenness, etc.); aircraft interior; damage; weapons; actions to overpower; police/fire rescue team assistance requested; demands; ethnic origin; physical description of person(s); political allegiances; ground services; airport installations; injuries/wounds; stowaways.

Re-routing/diversion Approach charts; procedures; routing; endurance; weather; airport installations; ground services; aircraft breakdowns.

Special flights ILS calibration; special test flight procedures; banners, balloons, etc.; ultralights, gliders; helicopters; aerial photography; highway watch; firefighting aircraft;

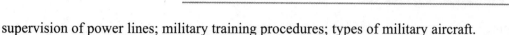

supervision of power lines; military training procedures; types of military aircraft.

Take-off incidents Abort; bird/animal hazards; traffic interference; runway incursion; overheating; towing; 180° turn back; runway excursion; cancellation and change of clearance; problems with steering gear, engine power; aircraft breakdown.

VFR flights lost/in difficulty Aerodrome environment; direction finder; manoeuvres for identification; endurance problems; installations at alternate/diversion field; forced/crash landing; ground services.

VIP flights Official ceremonies; protocol (greetings, etc.); ferry flight; military escort; diplomatic clearance; country names and nationalities; apron/ramp; terminal; boarding and disembarking of passengers; VIP vehicles; effects on traffic.

2. EVENTS AND DOMAINS LINKED TO EN-ROUTE AIR TRAFFIC CONTROL

Administrative problems Diplomatic clearances; customs regulations; civil service departments; impounded aircraft.

Aids for VFR flights Instrument panel; on-board equipment; pilot rating; flight plan; local place name; visual landmarks; positions; directions; endurance; aircraft breakdown; weather problems.

Aircraft breakdowns Instrument panel; instrument operation; radio beacon; positions/fixes; noises/sounds; smells; smoke; airport installations; ground services; engine performance; speed; relief/high ground; actions to solve problem; weather; fuel dumping/jettisoning; flight profile; structural damage (glass, metal); flight systems; aircraft controls; response to controls; airframe; warning lights; landing gear.

Aircraft proximity + pilot complaints Conflict situations; traffic load; aircraft characteristics; flight profile; weather conditions; injuries; distance/range; pilot manoeuvres; rules; procedures; avoiding action.

ATC system breakdowns ATC equipment/systems; radar display; radar performance; radio operation; previous messages; relaying messages; actions to repair; delays/duration; telephone lines.

Bomb scare Aircraft interior; search methods; fuel dumping/jettisoning; ground services; airport installations; ground movements.

Cargo problems/dangerous goods Packaging; substances; toxic substances; animals; smells; cabin equipment; load distribution; loading/unloading.

Change in flight plan Flight plan.

Collisions Airframe; structural damage (glass, metal, etc.); response to controls; debris; airport installations; ground services; relief/high ground; weather conditions; aerodynamic

behaviour.

Fire on board Outbreak of fire; control of fire; damage; aircraft interior.

Health problems Parts of the body; organs; symptoms; sicknesses; injuries/wounds; artificial limbs; medicine/drugs; first aid; medical equipment; medical staff; medical specialists; vaccines; quarantine.

Lack of fuel Airport facilities/installations; ground services; high ground; positions/locations; endurance/fuel remaining.

Misunderstandings Previous messages; types of messages; radio performance.

Passenger behaviour + unlawful interference Violent threatening behaviour; drugs; firearms; injuries; mental instability; nationalities; political allegiances; demands, threats; ground services; medical assistance; means of calming; means of overpowering; flight deck and cabin personnel.

Request to relay Names of people; means of relaying.

Special conditions on arrival State of the traffic on the ground; priority flights; industrial action; accidents; weather conditions on the ground; ground equipment failure; airport installations; ground services; curfew; approach procedures.

Special flights Type of aircraft; ferrying; diplomatic personnel; country names; nationalities; aeronautical military slang; military exercises; in-flight/mid-air refuelling; pilot manoeuvres; positions/fixes; weather conditions; VFR/IFR procedures; visual flight rules; airport installations; ground services.

Unauthorized manoeuvres Airspace; rules; previous messages; flight profile; positions/locations; stall levels.

Weather/MET problems Icing problems; clouds; struck by lightning; turbulence; external parts of aircraft; engine performance; response to controls; instrument performance; alarms; violent movements; relief/high ground; flight profile; injuries; blindness/loss of visibility.

3. OTHER DOMAINS

Activities on the field Change of runway and pattern; ramp vehicles; snow clearing; sweeping; mowing; harvesting; closure; opening of runway access roads; runway inspection.

Aerodrome/airfield environment Topography (hill, slope, coastline, forest, etc.); civil engineering (water, tower, bridge, pylon, etc.); high ground/terrain; built-up areas; roads and railway lines; power lines; cardinal points; particular local activities (firing range, etc.); agricultural activities.

Aircraft breakdowns Aircraft spare parts; systems (oxygen, hydraulic, electrical, de-icing, etc.); flight deck/cockpit; controls; instruments; instrument operation; noises and symptoms of malfunction; transponder problems; loss of radio contact; malfunctions; overheating (brakes, engine, etc.); fuel dumping/jettisoning; landing gear/tires.

Airfield facilities/installations ILS, radar, VOR, etc.; lighting systems; reliability of radio aids; direction finder; poor visibility equipment; aprons/tarmac/ramps; runways; taxiways; length and width of runway; parking zone; holding area; terminal; cargo area; bearing strength.

Ground services Opening hours; availability of services at night; assistance on the ground; safety altitude; passengers/persons on board; unserviceable equipment (stairs, luggage trolleys, etc.); auxiliary power unit; de-icing; refuelling; delay due to de-icing or refuelling; bird scaring; towing; firefighting methods; safety services; medical assistance; baggage handling.

Procedures Noise abatement; departure; approach; all-weather take-off and landing; go-around; holding procedures; land use; curfew; local residents.

References

[1] U.S. Department of Transportation Federal Aviation Administration. Airplane flying handbook[R/OL]. [2013-07-22] http://www.faa.gov/regulations policies/handbooks_manuals/ aircraft/airplane_handbook/media/faa-h-8083-3b.pdf.

[2] U.S. Department of Transportation Federal Aviation Administration. Aeronautical information manual [EB/OL]. [2013-08-22] Europe Comparison of ATM-Related OperationalPerformanc 2010. http://www.faa.gov/air_traffic/publications/.

[3] Airbus Industry. A320 flight operation manual[R/OL]. [2013-08-10] http://www.cockpitseeker. com/wp-content/uploads/A320/pdf/Print Only/PTM%20with%20airbus%20doc/pdf/ U0S2SP0-L. pdf.

[4] Anon. Airport and navigational lighting system aids[EB/OL]. [2013-07-20]http://www. Dauntless-soft.com/products/libarary/books/flt/chapter 14.

[5] Anon. Boeing company 777 flight crew operations manual-737NG.co.uk[R/OL]. [2013-07-24] http://www.737ng.co.uk/B777%20FCOM%20Flight%20Crew%20Operating%20Manual.

[6] U.S. Department of Transportation Federal Aviation Administration. Digital terminal procedures[R/OL]. [2013-05-27] http://aeronav.faa.gov/digital_tpp.asp?ver=1004&eff= 04-08- 2010&end=05-06-2010.

[7] International Civil Aviation Organization. 9432-2007 manual of radiotelephony[R/OL]. 4 th ed. [2013-08-12] http://www.freestd.us/soft2/605888.htm.

[8] International Civil Aviation Organization Radio Navigation Aids[M/OL]. 5 th ed. (1996-07-01) [2013-07-27]http://www.ihs.com/products/industry-standards/org/icao/historical/page8.aspx.

[9] U.S. Department of Transportation Federal Aviation Administration. Pilot's handbook of aeronautical knowledge，FAA-H-8083-25[M/OL]. [2013-07-26] http://www.faa.gov/regulations policies/handbooks_manuals/aviation/pilot_handbook.

[10] Anon. Incursion-avoidance[EB/OL]. [2013-08-03]http://www.flyngo.com/pilots-handbook-aero- nautical-knowledge-faa-h-8083-25a%20runway.

[11] U.S. Department of Transportation. Weight and balance, handbook 2007[M/OL]. [2013-08-11] http://www.faa.gov/regulations policies/handbooks manuals/aircraft/media/FAA-H-8083-1A.pdf.

[12] 教育部高等教育司. 大学英语课程教学要求[M]. 北京：外语教学与研究出版社，2004.

[13] 吴土星. 无线电陆空通话教程[M]. 北京：中国民航出版社，1996.

[14] 金霞. 新视野大学英语听说教程[M]. 北京：外语教学与研究出版社，2008.

[15] 郑树棠. 新视野大学英语阅读教程[M]. 北京：外语教学与研究出版社，2001.

[16] 李玉梅. 中国民航飞行人员英语阅读教程[M]. 北京：中国民航出版社，1997.